Introduction to Computer Aided Design, Engineering and Manufacturing

Using Mechanical Model

by
Prabhu Swaminathan
MSME (Mechanics)

 FriesenPress

One Printers Way
Altona, MB R0G 0B0
Canada

www.friesenpress.com

ISBN
978-1-03-830302-8 (Hardcover)
978-1-03-830301-1 (Paperback)
978-1-03-830303-5 (eBook)

1. COMPUTERS, SOFTWARE DEVELOPMENT & ENGINEERING, SYSTEMS ANALYSIS & DESIGN

Distributed to the trade by The Ingram Book Company

Figure 1. Model view of crane.

This project is model centric and is the focus point.

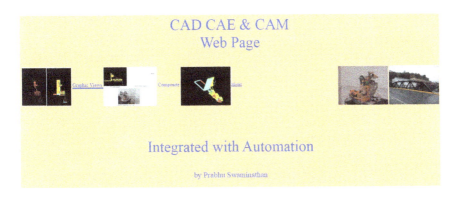

Figure 2. Image of web page.

Figure 3. Model view 2.

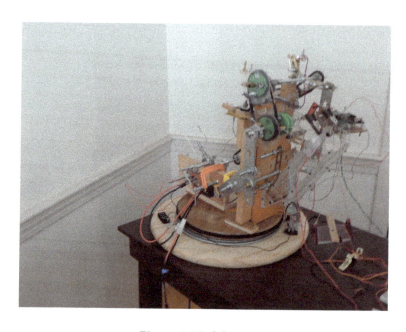

Figure 4. Model view 3.

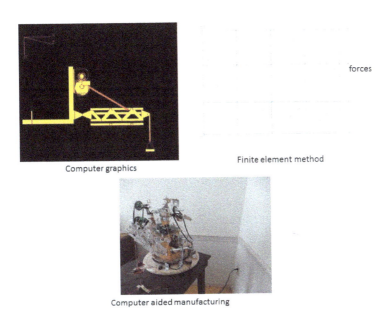

Computer graphics

forces

Finite element method

Computer aided manufacturing

Figure 5. A view of the complete process. Finite element model showing application of forces by arrows.

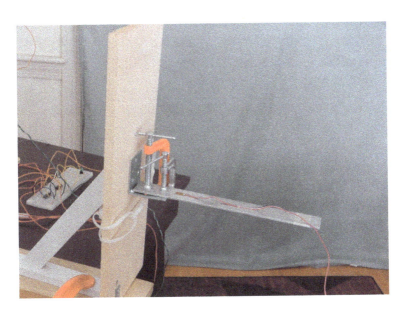

Figure 6. Strain gauge on an aluminum bar (experimental stress analysis).

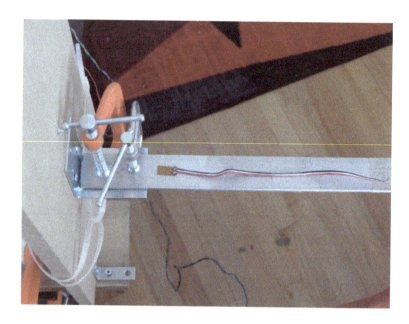

Figure 7. Top view of bar with strain gauge.

Figure 8. Possible field applications.

Disclaimer of Liability

The content and images in this book are for information only. The writer assumes no liability for omissions or errors in the content of this book.

Table of Contents

Tables

Figures

Preface

Computer-aided design (CAD), engineering (CAE), and manufacturing (CAM) are wide fields. Based on my hands-on experience, I bring these fields together into one unit. I add a physical model to enhance the explanation. It is maintained as educational tool.

After having worked at the companies listed below, I decided to document my experience, with the hope that it will be informative to younger students and young professionals starting their careers. Thus, with this intention, I have taken this step of writing this book, backing it up with my own educational and work experience.

I describe this subject matter and its application using a mechanical model. The mechanical model is a result of multiple activities described in the book. I highlight the basics of finite element analysis and experimental stress analysis with the model. The mechanical model operates wirelessly. The operational concepts can be projected to real prototypes for monitoring stress/strain characteristics of the system. This book is a blend of engineering and computer science disciplines.

As this is a self-created model, I decided to patent the model and related subject matter called Educational Training System (US Patent 11,893,902 B1, Feb 6, 2024). This training system touches on the fundamentals of CAD/CAE/CAM and its application using a mechanical model.

I was affiliated with these companies:
- General Dynamics Electric Boat, Connecticut: senior engineer
  - Company business: design and build submarines
- ASEC Engineering Corporation, Boston: senior structural engineer
  - Company business: design build structures, contributors to the Big Dig project in Boston

- Badger Engineering Corporation, Boston: senior structural engineer
    - Company business: design and engineer petro chemical plants
- Electronic Data Systems, Detroit: engineer
    - Company business: support for automotive company
- Stone and Webster Eng. Corporation, Boston: structural engineer
    - Company business: design and build power plants
- Charles T. Main, Boston: junior structural engineer
    - Power company: design and build transmission lines/towers

My education:
- MSME (Mechanics) Northeastern University, Boston, MA
    - Thesis: finite element analysis for plate in plane stress/strain and plate bending
- MSCE, Northeastern University, Boston, MA

Introduction

This manual is an introduction to computer-aided design (CAD) computer-aided engineering (CAE), and computer-aided manufacturing (CAM) and addresses the fundamentals in applied mechanics and applied computer sciences. It highlights a concept of runtime experimental stress analysis (ESA), a runtime theoretical finite element analysis (FEA) with graphic simulation, a graphical user interface (GUI) with Java and Common Object Request Broker Architecture (CORBA), and a CAM-generated model from the outline developed in graphics. Thus, it builds a concept in prefabrication and post-fabrication stress analysis from the model. Pooling all of these players in an integrated fashion, along with the mechanical model's wireless operation and the resulting concept from collecting runtime data from post-fabrication, make this manual different in its field. A physical model, demonstrated throughout this manual, ties all these players together.

In writing this manual I have drawn concepts from my hands-on experience and research from working in a variety of industries and studying at educational institutes. These applications include structural and software discipline associated with transmission lines (towers), nuclear power plant, petro chemical plant, auto industry, tunnels (Big Dig), and ship-building (submarine) industries. Thus, it is an outcome from a collection of segmental experiences derived from a variety of industries. The model is also outcome from several classes at Northeastern University (with a finite element analysis thesis) and Harvard Extension school classes, where it was initially developed via the Gibbs CAM tool. Consequently, some of the presentations are from my class notes. This model that I created by myself is an example of multi-phase development.

There are 10 parts to this manual, each introducing the relevant concept with the model as a backdrop. I keep our focus on building a CAM model and describe all the steps to accomplish this small task. I've kept the explanation simple as this is meant to be introductory. The model in figure 1 is an example of outgrowth from this study. It is fabricated to be a table-top display.

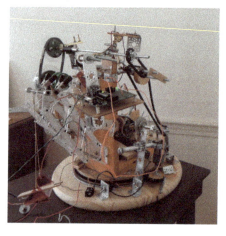

All the programs accompanying the model are saved on thumb drives and can be executed from the instruction in the appendix. The website links to a demonstration with the model. The thumb drive also includes a recording of remote operation.

The first two parts of this manual are on software and engineering respectively. Software engineering is a system comprising of various entities and maintaining the end product. This software highlights the entities shown in the diagram.

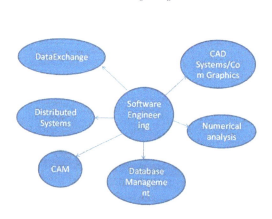

Figure 9. Entities (displayed in blue).

The entities that this document addresses are shown above in figure 9. Keeping this as our focus, we will develop a product with Java as our development kit.

Java/CORBA, .NET, XML, HTML, Computer graphics, CAE, and CAM are each vast subjects. It will be impossible to present this vast field in a single document such as this one; therefore, only the concepts are drawn from each of these fields and the manual is a collection of various concepts working together. We develop an introductory CAD and CAE and use Gibbs CAM software for fabricating the model.

In the field of manufacturing there are several participants. The initial step to connect these participants is a CAD model. These participants are tied to different platforms of computer systems. Since participants must communicate with each other, there is a need to develop a method that could be understood between all the manufacturers, or agents. This task is achieved by forming neutral files of a CAD model. Neutral file is one way of keeping the CAD models platform independent. Initial Graphics Exchange Specification (IGES) and Standard for the Exchange of Product Model Data (STEP) are two types of formats that are popular in the CAD world and are the current methods of forming neutral files. Yet another method is direct transfer. This is a collaborative effort between various CAD systems. There are also other methods, but generally all commercial CAD systems are equipped to understand these formats except the direct transfer.

In this manual, we will also highlight features from database programming. Storage and retrieval of information is an integral part of this process. We chose Java as our development language. An attempt is made to bring all these features together and present them as a source of reference. We start by applying all the above to a case of rapid prototyping and focus on developing a model from the variety of concepts described here. Ours will be a model with experimental phase attached to it. Our purpose will therefore be to add an experimental verification of theory.

There are several books now in place on FEA and computer graphics, so why this one? Whereas the other books have been written by top-notch authors, this manual is of practical nature and brings the process of integration among most of the systems together (i.e., computer systems, database management, FEA, boundary element analysis [BEA], experimental stress analysis [ESA] data translation and manufacturing). All demonstration with respect to the model is different as well.

The section on computer graphics forms a foundation in building a mathematical model in CAD, allowing it to turn it into a physical model. Mathematics, however, have not been reproduced. Instead, pointers are given in the bibliography where mathematical details are found; however, software implements this necessary mathematics.

The ESA section creates a path to verify experimentally theoretical values from stress analysis. This interfaces with **LabVIEW** for monitoring variation of stresses. (LabVIEW is also used for strain measurement.) A separate model using potentiometer and using strain gauge on a beam model is demonstrated. Subsequently, the concept is projected to the crane model. Experimental stress analysis is my own contraption of adding devices to supplement the methods that exist, such as adding a contraption with MicroStrain and LabVIEW software and hardware. Here I dwell on the behavior of the model using these contraptions.

I have also added a section for acquiring the strain data wirelessly from a remote device provided by LORD MicroStrain company and projected its application to the model and real-world problems. Coupled with the remote device, this introduces a concept of monitoring the status of stress in the crane boom continuously as it takes up various positions in the vertical plane, also allowing variable loads. A demonstration is included in the thumb drive.

Some other key aspects:
- This book's content is arranged sequentially, following the steps in the flow diagram shown on the rapid prototyping page.
- The experimental postfabrication is an example of capturing postrun-time behavior of the model through stress analysis methodology.
- The model's outline is based on prefabrication stress analysis concept. Thus, prefabrication and postfabrication stress analysis are illustrated.
- Vibrations are addressed because of their application with seismic actions on pipe racks of nuclear power plant. A section on

- **orthogonal** vectors is included because this is an alternate method to determine dominant eigenvalues for the vibrations.
- On the issue of network, I present a simple client-server example with CORBA. This adds a concept in networks and maintains an integrated approach with GUI.
- Animation in graphics is coupled with stress analysis of the crane boom component of the model for selected positions in the vertical plane.
- Introduction to finite element analysis is included as a CAE part to demonstrate its application on the model.
- Database is a general-purpose storage and retrieval method. I present a simple database structure for storing member properties from American Institute of Steel Construction (AISC) tables that can be used for the finite element method.
- The topic of boundary element is addressed in a more cursory way. This is to cover the potential of numerical analysis (CAE) in solid mechanics. Pointers are given to this method that adds to the strength of finite element method.

Key point: these features are projected into a scaled-down physical model. To broaden the scope, they are projected to real life projects as a possible application.

This manual is a result of survey from the material in bibliography. It is intended to be an integrated tool and an introduction to the subject of CAD/CAE/CAM in a different form. Software was built using Java and FORTRAN programs. The version on Java was jdk1.8.0_05, jre1.8.0_171; the version on FORTRAN was G77; and MySQL version 1.0.

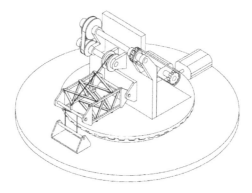

Figure 10. Outline of model.

Proofs are documented with screenshots from computer output. Upon request, the writer can provide a thumb drive where workings can be seen.

About Software Package

The thumb drive labeled "HTMLDemo" has examples of automation and workings of the model. This package is a collection of all executables pertaining to the model. These are for demonstration purposes only and are meant to support the manual.

All the programs are listed here. User must run the following initialization:

Open a command shell c:>
Change directory to C::\users\Admin> will need to make a batch file, setup2 (contains all paths)
(See appendix C for a screen capture.)

Rule 1: For compiling Java programs follow these listed steps -
Insert thumb drive labeled Java into a USB port.
Change the drive letter to D:>.
Move into fnprj folder by CD:> fnprj
Run batch file
1-cmpl1 (for Assembly.java) (for merging objects)
2-cmpl2 (for Diagram.java) (for displaying view)
3-cmpl3 (for Annimat1.java) (for animation)
4-cmpl4 (for Model3D.java) (for converting 3D to 2D after transformation and scaling)

Rule 2: follow initialization as above
For compiling Fortran programs go to C: drive and enter "/F/G77" directory
Type "g77setup", this sets up the path. (See Appendix C.)

Insert the thumb drive labeled "Fortran" into second USB port
Rename this drive "F"
From F::> drive
Move to directory "fortrancom /fort99/g77"
Then compile with "ccfor" batch file on directory
"fortrancom /fort99/g77"

Rule 3: For the Fortran programs to run, the drive on the thumb drive and ProcessBuild must be compatible. If the drive letter on thumb drive changes, then the drive letter in ProcessBuild.java must change and recompiled.

The following are located on the thumb drives:
1. Client/server
2. Java programs
3. Fortran programs

Two USB ports are used: one defined as D; the second as F.

Definition: cd for changing directory
F:> path contains Fortran programs
F: cd fortancom/ fortrancom/ fort99/ g77
Batch file "ccfor" compiles Fortran programs
D:> drive contains java programs

Computer graphics program (develops outline of the model and other functions from GUI)

The following java programs are compiled to java class:
- **Assembly.java** merges the file from object.dat and saves as 3D file called output2.dat. Also integrates to finite element programs.
- **Model3d.java** reads the output2.dat 3D file from Assembly; performs object orientation, perspective view implementation, and scaling; and saves as 2D file called product_file.dat.

Limitations:
- Node coordinates must be sequentially listed in output2.dat.

- **Diagram.java** reads the product_file.dat from above. This program for displaying the model as a wireframe, shading, and rotations. It has a button for animation.
- **Annimat1.java** performs animation and writes to a file for different angles of rotation and stress analysis of truss.

For animation to work, both the thumb drives—one containing Assembly and another containing Fortran programs—must be inserted in the USB port. If the Fortran thumb drive is not inserted, then animation from Assembly will not work because the animation module is designed to write files into the Fortran program files. If it does not see the Fortran thumb drive when animation is executed, it will give a null pointer exception error. Fortran programs are listed in Assembly.java display screen.

Batch File Operations

@promt D>Type "CADCAM" followed with sequential returns @prompts to CADCAMCAE workflow screen (described in appendix C).
The following topics accompany this project:
- Java class code for CORBA client-server initiation and initiation of graphic model through workflow GUI.
- Executable code for finite element program as applicable to the model.
- Executable code for eigenvalue and eigenvector in support of finite element method.
- Description of IGES and STEP methodology of data translation. IGES methodology: The reader may reference to commercial software cited in bibliography for an in-depth application of IGES methodology
- Fabricated CAM model.
- Pointer to experimental stress analysis using strain gauge on the model but exercised outside of model with potentiometer and connection with LabVIEW.
- Pointer to remote collection of data on strains via
- MicroStrain devices.

- Demonstration of workings of model through radio frequency controls on thumb drive.

List of Fortran Batch Files

Noted here are batch files. References are cited in appendix B for these programs.

Beams.bat
beam2 (develops stiffness matrix)
sprngk (adds spring constants)
load (develops load vector)
udl (develops uniformly distributed load vector)
disp (applies displacement to input)
rdload2 (reform load vector based on end conditions)
bound (applies boundary conditions)
bsolve1 (solves the matrix)
resultant (computes resultant at boundaries)
elreac (results with elastic spring constants)
beamfrce (computes beam end forces)
cccccccccccccccccccccccccccccccccccccccccc

This is a revised batch file for "beam," reducing the beams.bat to a shortened batch program. It was typically chosen to compare FEA results with experimental strain gauge setup.

beams2.bat
beam2
load
udl
disp
bound
bsolve1
resultnt
beamfrce
ccccccccccccccccccccccccccccccc

Plane truss

Pltruss.bat or ptruss.bat

PlTruss.for

load

bound

bsolve1

resultnt

ptrussfrce

mendtrussfrce

added space truss batch file

sptruss2.bat

cccccccccccccccccccccccccccccc

Plane frame (pframe.bat)

Plframe

Load

bound

bsolve1

resultnt

pframefrce

mendframefrce

ccccccccccccccccccccccccccccccccc

Plane grid

plgrid

load

bound

bsolve1

resultnt

gridfrce

gridendfrce

ccccccccccccccccccccccccccccccccccc

Space truss (batch sptruss2.bat)

PlTruss.for (generates planer truss coordinates at run time)

GenDataSpTruss (generates space truss joint coordinates)

GenNewSptruss (reads data from GenDataSpTruss)

sptrus2

load
bound
bsolve1
resultnt
sptrussfrce
sptmemendfrce
ccccccccccccccccccccccccccccccccccccc

Space frame
// to generate opposite side node from plane truss simulation
// of the model geometry

GenSpData
GenNewSpMast2
mast1
frmld1
bound
bsolve1
resultnt
framememberfrce (frame member end forces)
frameGLendfrce (global end forces)
ccccccccccccccccccccccccccccc

SpaceFrame
mast.bat
mast1
frmld1
frmbnd
bound
bsolve1
cccccccccccccccccccccccccccccc

Plane stress rectangular element((plstressR.bat)
rectpl
load (force7.dat)
bound
bsolve1

rctplst

cccccccccccccccccccccccccccccccccc

Plane stress triangular element(plstressT.bat)

Tripl (**nodedataT.dat**)

Load (force8.dat)

bound

bsolve1

triplnstress

ccccccccccccccccccccccccccccccc

Plane stress triangular element and plane truss combination

Tripltruss.bat

tripl_d

load

bound

bsolve1

resultnt

cccccccccccccccccccccccccccccccccc

Plate bending rectangular element (plbend1.bat)

plbendr

load

bound

bsolve1

plbendStrsR

ccccccccccccccccccccccccccccccccccccc

Plate bending triangular element (plbndT.bat)

plbendT1 (nodedataT.dat)

load (force10.dat)

bound

bsolve1

tplstrs

cccccccccccccccccccccccccccccccccccccccc

Orthogonal vectors (beamsOr.bat)

beam2

load

udl

rdload2
bound
bsolve2
ccccccccccccccccccccccccccccccccccccccc

Beam dynamics (beamd.bat)
beam2
load
disp
rdload2
bound
cndnse
matinv
dyan2_1.exe (input cndnse.mass, for Eigen values and vectors)
ccccccccccccccccccccccccccccccc
For boundary elements, interested readers may want to refer to cited references in bibliography.
ccccccccccccccccccccccccccccccccc

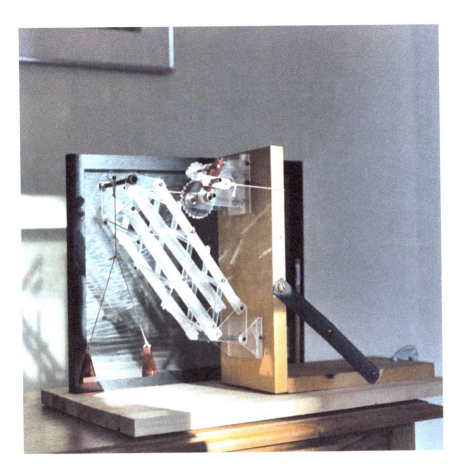

Figure 11. CAM model.

Scope of Project

This is a small software/engineering project in Java. It integrates software, LabVIEW data, MicroStrain data, graphics, FEA, and HTML, and is thus a self-contained unit with a working model. The model was started at Harvard Extension School and was enhanced further with an objective to introduce the concept as an educational tool.

A concept of remotely collecting strain data is addressed through this model at various stages of movement of the crane arm. This project uses Java, and Fortran languages for development. Fortran programs are executed from Java graphical user interface. An attempt is made to collect data through remote receiver device (wireless communication).

The difference between the current approach and other existing software is that this software leads to a manufacturing path and is a self-contained unit of introduction.

Workings of this manual can be seen by visiting the thumb drive tagged as "HTMLDemo."

Objective

Design and develop a small software/engineering project demonstrating integration of various entities of software. These features will introduce networking, computer graphics, finite element analysis, database systems, data exchange, and CAM.

Fabricate a model using CAM tools (From GibbsCAM). Our CAM development is a two-part development. The first part being to fabricate a model followed by a second part to add automation and introduce stress analysis features.

Emphasis is on building a compact, integrated system including automation. We will build a framework for automation using the CAM technology with radio frequency (RF) control.

Simulate stresses in the system using theoretical (FEM) and experimental approach in LabVIEW.

Network using distributed objects such as CORBA. Software will have graphical user interface for segmented actions and display. HTML is used for web page development and linked with demonstrations of automation. This project draws concepts from different disciplines and is thus an interdisciplinary activity. This will have an RF interface attached to it that allows operation of the model using a remote controller. This operation can be viewed on the HTML page.

PART 1:
Introduction

Rapid Prototyping

Rapid prototyping is a methodology of prototyping parts in quick succession. This effort helps in verifying a variety of factors, such as the size of actual part or studying the tolerance between components, in a tight knit space. There are several methods of prototyping the parts. We take a simplified approach of using a typical CAM machine (GibbsCam) and a scaled-down version for producing parts. Although this has its own platform, it is integrated with two other process called CAD and CAE. CAD geometry is the basic entity for manufacturing. We will touch on the fundamentals that make this whole process work together.

The standard file format for this task is a .STL file. A general flow path to manufacturing a product is shown in figure 12. Usually, this feature takes place via data translation described later in this manual.

Participants in Manufacturing

There are several participants in manufacturing in current times. Figure 12 shows examples as well as typical activities in the industry. Typically, a manufacturing process involves all the activities shown here. (From writers article on data translation, cited in bibliography)

How Do We Manufacture a Product?

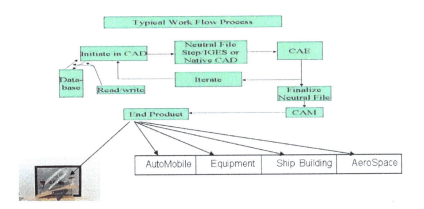

Figure 12. CAD/CAE/CAM workflow 1.

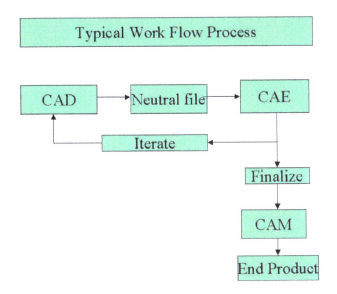

Figure 13. Typical workflow 2.

We walk through each of these, detailing the necessary steps only. The tasks are split into the following segments:

1. Build a small network with CORBA as the client-server interface. Client will send request to the server and server will respond. In turn, the server will invoke the database. In this case, Mysql.cfg. (This is not in current software, but it can be added on if the reader is interested to enhance.)

2. Build a database system that will store elements of AISC tables.

3. Make an outline of a model using concepts from computer graphics.

4. Explore basics of finite elements, as this is used by industries, with application to the model.

5. Explore basics of photo elasticity and an alternative methodology by using strain gauges with interface to LabVIEW, to supplement verification of theoretical results.

6. Add a wireless acquisition of data as a run-time concept from the model.

7. Explore data translation foundations and apply that to manufacturing via GibsCAM tools.

8. Subsequently build the model and simulate the movement with an RF interface.

9. Explore the strain conditions under movement or in motion of the model.

PART 2:
Typical Integrated Computer Environment

Background

A typical integrated work environment for processing activities is built from the following items shown in figure 14 below. This approach is used to build an integrated software system using concepts from distributed objects. This integrates using GUI, a model display, finite element examples, and animation on graphics model, coupled with stress analysis on a component of model. This serves as a client-server model.

This is an example of code reuse, where applications are developed for alternate usage from base line software.

Server

The function of a server is to bind to the port and listen to the client, then give response to the client based on demand. A server is accessed via one of the remote object access methods. We chose to perform this task with Java and CORBA because these two mix well together. Other methods of communication are with Unix sockets and remote monitoring interface (RMI). CORBA methodology is used in this manual.

Figure 14 shows a typical computer interaction model listing various activities. We will highlight each individual activity and elaborate on the necessary connections individually. If MySQL database server is downloaded, then this connection can be made to the database. Other functions of the database will then be exercised on the database. Some of them are listed in "Data Base" (DB) below. Interested readers can use that option for storing parts and so forth.

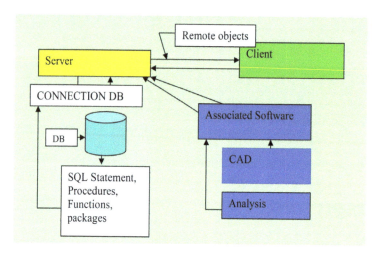

Figure 14. Image of an integrated system. Typical computer environment setup.

Communications

Our objective is to highlight features of communication in CORBA, a vast and intricate element. We demonstrate an integration of CAD/CAE/CAM with CORBA on a small computer graphics program created in Java. We adopt a scaled down version from this "distributed system" technology and add it to our crane model as an application. Although a unit of client-server, we leverage this method in building a GUI. This serves as a driver to integrate other function with a graphical user interface. Using COBRA, we build a standalone server on a local computer. We then communicate with a client and execute the functions from a workflow management GUI.

The images in the next few pages are screen shots captured from this operation. MySQL is a database software that can be integrated with CORBA.

Using local port number and IP address are the key elements for the server unit that allows a user to get access to functions on the server. Java has two methodologies: RMI and CORBA. These are communication methodologies to implement on a server using concepts from distributing objects.

The images or screenshots below show the options on the server on a local computer.

Typical Screen Images

WELCOME   Server Started

Figure 15. Image of server page.

Workflow screen

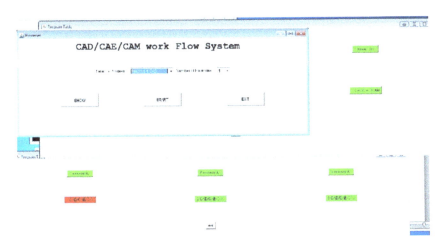

Figure 16. Workflow image in front and program selection option table in background. (Red button indicative of option selected.)

Figure 17. Start activates the screen-showing option for selecting a program.

Building a Server Using CORBA[1]

CORBA methodology is built from few different players. To accomplish this integration, they are summarized below as components of CORBA. For more details, reference material is cited in the bibliography.

- Interface definition language (IDL) to build an interface
- Object request broker (ORB)
- Stub (client side) and Skeleton (server side)
- Factory
- Servant
- Naming services
- Event messaging services
- Consumer admin
- Supplier admin
- Notification methods (using event channel)
- Application to connect a MySQL database (for future)

We demonstrate a standalone server. We make use of Java remote object services, transient naming service **(tnamserv).** This is a utility provided

---

1. Jim Farley, *Java Distributed Computing* (Sebastopol, California: O'Reilly Media, 1998), 58.

by Sun Microsystems. There are a few different players to accomplish this integration. We built the server using **CORBA ORB** and **event services** and implemented **PUSH** model on the event channel.[2] ORB is the object request broker and a CORBA communication channel through which all the transactions take place. Event services is a service provided by CORBA for communication between consumers and the suppliers, whereas ORB is a part of an infrastructure and transfers the data. The event channel is used for notification. The mechanics of the event channel is an additional communication channel that is provided by the object management group (OMG) and uses another level of communication for notification.[3] Typical event channel flow is shown in the image below (figure 18). This is a bidirectional flow of events.

There are two distinct styles of communication: a push model and a pull model. Event channel is a communication channel between the supplier and consumer. Agents (supplier and consumer) register with an event channel (see footnote 3).

Elements of CORBA procedures:

IDL to Java language
CORBA event services
ORB technology

The following are the modules created by the OMG. Some of the terms are recaptured here (for detailed explanations, see footnote 3):

Consumer admin
Supplier admin
Push supplier
Proxy push supplier
Push consumer
Proxy push consumer
Event channel

---

2.   See appendix B of the reference above.", Event Service- propagation model of event services 321

3.   Farley, *Java Distributed Computing*. appendix B 318,321

Interface definition language to Java is a conversion language that builds the interface in Java, including client side stub and server side skeleton. The interface has the name of the methods and variables. The syntax for using this **idltojava** compiler is shown in table 1. **Myfile.idl** is used (pseudonym). For its data structure, see reference above.

Table 1. IDL.

Function name	Argument 1	**Argument 2**
@promt c>Idltojava	-fno-cpp	**Myfile.idl**

The result of this action will build an interface.java file.

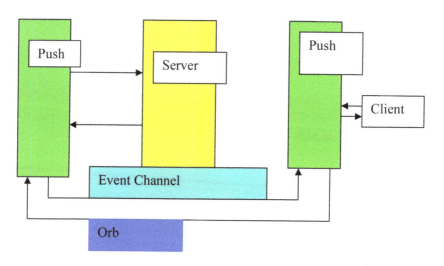

Figure 18. Diagram of bidirectional ORB and event channel with push and pull model.

Typical images of client-server

Typical images from client-server are shown on previous page (figures 15 to 18). This setup is adopted for integrating other functionalities, such as graphics display of model and finite element programs.

MySQL Database

Background

A database is always a part of an integrated system. Interested readers may want to explore this as an add on.

MySQL is a free software and can be downloaded from the website. It is not the intent to scope out the entire MySQL; instead, we highlight the concept and its use. The common approach to this is using the sequel query language (SQL). Typically, this option can be applied to creating a part list of the model with IDs assigned to each part. This can be linked with CORBA application as an option. It is left out to conserve the scope of the subject.

Member Properties

We save American Institute for Steel Construction (AISC) member properties in XML format for use with finite element application.

We initially load up the table in **XML** format. We use **Xerces Dom parser** to read the file and display its contents. Xerces Document parser is also available as free software.

We store the data from AISC[4] table in XML format for **structural engineering** application. This connects with the CAE part of this manual by adding a feature to analyze a prototype type after noticing a satisfactory performance of the model. If the model is scaled up to be a prototype in **steel**, these members properties can be used by same finite element program described later in the section. The program can be enhanced to select member properties based on constraint established from finite element analysis (i.e., using Pass/Fail option). This program uses predefined properties only. Prototype can be subjected to wireless acquisition of data described later in experimental stress analysis.

Listing in XML format is displayed.

---

4.   American Institute of Steel Construction, *Steel Construction Manual*, 8th ed. (Chicago, Illinois: American Institute of Steel Construction, 1980).pages 1-14 to 1-46

Viewing Database Tables

This is normally covered in the manual of MySQL database. Typically, these are SQL query language statements.

Member properties are arranged in the following format (these are extracted from AISC manual): see footnote 4. "AiscTable.dtd" is a file and defines the tagged item using regular expression.

```
<?xml version="1.0" encoding="UTF-8" standalone="no"?>
<!DOCTYPE Aisc SYSTEM "AiscTable.dtd"><Aisc>
<Title>Member properties</Title>
<Title>Source</Title>
<Title>AISC</Title>
<Elasticity>29</Elasticity>
<PropertyData>
<name>W36x300</name><area>88.3</area><Ix>20300</
Ix><Iy>1300</Iy>
</PropertyData>
<PropertyData>
<name>W36x280</name><area>82.4</area><Ix>18900</
Ix><Iy>1200</Iy>
<PropertyData>
<name>W36x280</name><area>82.4</area><Ix>18900</
Ix><Iy>1200</Iy>
</PropertyData>
<PropertyData>
<name>W36x260</name><area>76.5</area><Ix>17300</
Ix><Iy>1090</Iy>
</PropertyData>
<PropertyData>
<name>W36x245</name><area>72.1</area><Ix>16100</
Ix><Iy>1010</Iy>
</PropertyData>
<PropertyData><name>W36x230</name><area>67.6</
area><Ix>15000</Ix><Iy>940</Iy>
</PropertyData>
```

```
<PropertyData><name>W36x210</name><area>61.8</area><Ix>13200</Ix><Iy>411</Iy>
</PropertyData>
<PropertyData><name>W36x194</name><area>57.6</area><Ix>12100</Ix><Iy>375</Iy>
</PropertyData>
<PropertyData><name>W36x182</name><area>53.6</area><Ix>11300</Ix><Iy>347</Iy>
</PropertyData>
<PropertyData><name>W36x170</name><area>50.0</area><Ix>10500</Ix><Iy>320</Iy>
</PropertyData>
<PropertyData><name>W36x160</name><area>47.0</area><Ix>9750</Ix><Iy>295</Iy>
</PropertyData>
<PropertyData><name>W36x150</name><area>44.2</area><Ix>9040</Ix><Iy>276</Iy>
</PropertyData>
<PropertyData><name>W36x135</name><area>39.7</area><Ix>7800</Ix><Iy>225</Iy>
</PropertyData>
<PropertyData><name>W33x241</name><area>70.9</area><Ix>14200</Ix><Iy>932</Iy>
</PropertyData>
<PropertyData><name>W33x221</name><area>65.0</area><Ix>12800</Ix><Iy>840</Iy>
</PropertyData>
<PropertyData><name>W33x201</name><area>59.1</area><Ix>11500</Ix><Iy>749</Iy>
</PropertyData>
<PropertyData><name>W33x152</name><area>44.7</area><Ix>8160</Ix><Iy>273</Iy>
</PropertyData>
<PropertyData><name>W33x141</name><area>41.6</area><Ix>7450</Ix><Iy>246</Iy>
```

```
</PropertyData>
<PropertyData><name>W33x130</name><area>38.3</
area><Ix>6710</Ix><Iy>218</Iy>
</PropertyData>
<PropertyData><name>W33x118</name><area>34.7</
area><Ix>5900</Ix><Iy>187</Iy>
</PropertyData>
<PropertyData><name>W30x211</name><area>62.0</
area><Ix>10300</Ix><Iy>663</Iy>
</PropertyData>
<PropertyData><name>W30x191</name><area>56.1</
area><Ix>9170</Ix><Iy>598</Iy>
</PropertyData>
<PropertyData><name>W30x173</name><area>50.8</
area><Ix>8200</Ix><Iy>539</Iy>
</PropertyData>
<PropertyData><name>W30x132</name><area>38.9</
area><Ix>5770</Ix><Iy>380</Iy>
</PropertyData>
<PropertyData><name>W30x124</name><area>36.5</
area><Ix>5360</Ix><Iy>355</Iy>
</PropertyData>
...

</Aisc>
```
End of table

Conclusion

In this chapter, we introduced a concept of networking using a distributed system. We utilized CORBA and an associated communication channel. A concept of creating a parts list using MySQL database for the model was introduced. We also created screens of graphical user interface in Java allowing other programs (graphics of model and Fortran) to be exercised from the graphics screen. We also added an XML format for storing the properties of AISC tables for ready reference, allowing for selection at a later time.

PART 3:
Computer Graphics: Framework

Background.[5]

Computer graphics is a digital methodology of presenting the objects that we perceive with our eyes. It has the means of displaying the spatial arrangements and spacing between the objects. It greatly facilitates identifying the clearances and conflicts between various objects that can be quickly resolved. Its foundation lies in describing objects in mathematical terms and some of that is listed below in the table 2. Industry is a complex system of objects and requires that the object be placed without any interference. Model development is presented below as an outline.

Objective

Our first objective is to develop an outline of a model using graphic tools developed here and demonstrate animation, coupled with FEA.

Subsequently, fabricate a model matching the outline and using CAM tools, introduce remote operations on the model, and perform stress analysis on the model. This concept is demonstrated on a beam model in

---

5.   See David F. Rogers and J. Alan Adams, *Mathematical Elements in Computer Graphics*; Weiskamp et.al, "Power Graphics"; Ian O. Angell and Gareth Griffith, *High-Resolution Computer Graphics Using FORTRAN 77.*

the Experimental Stress Analysis section. We will present graphics using different viewing options of this outline.

Integration flow

Flow path of Integration

Representation of Flow Path

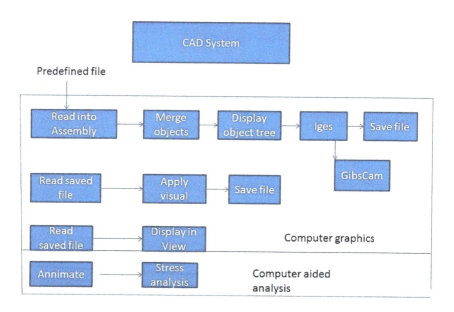

Figure 19. Adopted flow path.

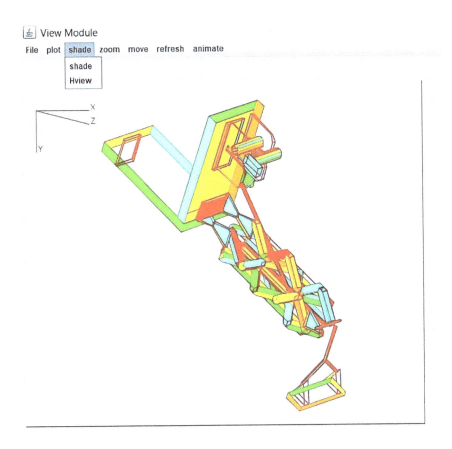

Figure 20. Image of complete model.

Data translation. Data translation is integrated in this connection (ref. rapid prototyping page). The graphic model exhibits animation and has its application projected onto the model in conjunction with numerical analysis (CAE). The animation part couples with finite element analysis of the boom. The model itself is self-designed and expresses an independent thought in computer graphics and its use. While there are numerous books on computer graphics, our example is different in the sense that we create an outline of a model using simplified graphic elements and finally fabricate a physical model that closely resembles the graphic model with a CAM tool. Our model is a model of a crane.

Animation. The animated part of the graphics model will write to files depending upon the movement of the boom. A run-time simulation of finite element analysis is coupled with the movement of model crane arm. These files can later be developed to real-world coordinates and be subjected to stress analysis.

Mathematics of graphic model. The mathematics portion of the graphic model uses features from coordinate geometry such as planes, normal, vector dot product, vector cross product, matrices, and intersection of lines. We use an algorithm of tracing points "inside and out" on a bounded region for shading. This segment is exhibited with self-developed software in Java.

Structure for a complete display. The software
1. Reads a predefined file (object.dat) and assembles the product in Assembly.java to make a complete model (see figure 20).
2. Reads the saved file from one (called output2.dat) and applies transformation and scaling in model3d.java and saves the model as product_file.dat.
3. Reads the file from two and displays the model in Diagram.java.
4. Reads the file from two and displays animation and creates new file in animate.java. While in animation, component of a model (i.e., truss or boom) is analyzed for stress with interface to Fortran program. This is an example of run-time simulation.

An example is included that displays minimal CAD system. Procedure is described in appendix C. This includes a display of stored objects, a tree display, wireframe, and a shaded display. The model development is made up of the following sequences:

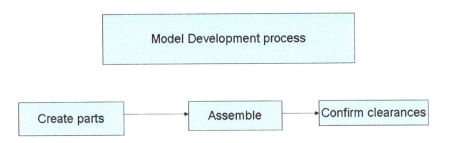

Figure 21. Typical steps in development.

Sequence 1: analytical geometry (mathematics) typically surveyed for the development of graphics model.[6] This segment uses concepts from coordinate geometry. These elements of coordinate geometry are listed in table 2.

Analytic geometry is made up from the following elements:

Parametric equation of line
Properties of lines and intersection of lines
Directed line segment
Direction cosine of a line
Linear transformations
Properties of plane
Normal from determinants
Dot product and cross product
Intersection of line and plane
Distance of plane from a fixed point
Cubic spline
Space curves such as B-spline
Rational B-spline
B-spline, rational B-spline surfaces

---

6.   See: David F. Rogers and J. Alan Adams; Ian O. Angell and Gareth Griffith; Alan Watt; and James Stewart.

Sequence 2: coordinate transformations and display. For an angle defined by theta, program uses the transformation in two dimensions and is expressed by equation 1.

$$\begin{bmatrix} Cos\ (theta) & Sin\ (theta) \\ -Sin\ (theta) & Cos\ (theta) \end{bmatrix}$$ Equation 1

Figure 22. Coordinate transformation

Sequence 3: animation and simultaneous stress analysis. A graphical model is created using the simple geometrical properties described here. The graphical model is transformed into a manufactured product that is linked with the other mechanical operations (i.e., automation, experimental stress analysis, and simulation by movement [animation] of the model arm). The graphics display of model is updated with movement of arm, with colors changing for each movement of the crane arm. We integrate Java with Fortran programs in this sequence.

Some Geometric Entities and Common Operations

Some of the geometric entities are added here. These typically make up a basic graphics package. Details can be referred to computer graphics books listed in the footnote and in the bibliography. Some excerpts are addressed here.

Directed Line Segment[7]

A directed line segment from point P1 to P2 is represented by:

$$P(x) = P1U + (1-U) P2 \qquad\qquad \text{Equation 2}$$

Where U = fractional distance from P1.
If U = x/L , 1- U = 1- x/L
L = length of line segment.
P(x) = P1(x/L) + (1-x/L)P2
At x=0; P(0) = P2;
At x= L (length of line segment), P(L) = P1
Then this equation gives a linear variation between P1 and P2. X is a distance from
P1 to P2.
Where: P1 is defined as points X1, Y1, and Z1
          Point P2 is defined with X2, Y2, and Z2

Intersection of Line and Plane[8] external source

Plane: equation of plane

Equation of a plane is given by:

$$A*Xn + B*Yn + C*Zn + D = 0. \qquad\qquad \text{Equation 3}$$

Xn, Yn, and Zn are coordinates of three points.
A, B, and C are defined as normal of a plane. D is the distance of the current plane from a reference plane.

---

7. Ion O.Angelli and Gareth Griffith, 128 w/permission
    David F. Rogers and J. Alan Adams, *Mathematical Elements in Computer Graphics* (New York: McGraw-Hill, 1989), 211.

8. Ian O. Angell and Gareth Griffith, *High-Resolution Computer Graphics Using FORTRAN 77* (New York: Palgrave, 1987), 113.

We compute A, B, C, and D constants. There are four constants so we must have four equations to find the values. We solve the determinant for getting the normal from selected points (X0,Y0,Z0), (X1, Y1, Z1) and (X2, Y2, Z2).

If D is set to zero or the origin, then we have three unknowns.

We pass the parameters of two direction vectors computed from spatial coordinates (X0,Y0,Z0), (X1, Y1, Z1), (X2, Y2, Z2) and solve the determinant.

A, B, and C are determined from the cross product..

dX1= X1-X0; dY1= Y1-Y0; dZ1= Z1-Z0;

dX2 = X2- X0; dY2 = Y2-Y0; dZ2 = Z2 - Z0;

where dX(1,2), dY(1,2), dZ(1,2) are directional vectors

$$\begin{vmatrix} i & j & k \\ dX1 & dY1 & dZ1 \\ dX2 & dY2 & dZ2 \end{vmatrix}$$

I=A = dY1*dZ2 – dY2*dZ1

J=B = dX1*dZ2 – dZ1*dX2

K=C = dX1*dY2 – dY1*dX2

Thus A(X-X0) + B(Y-Y0) + C(Z-Z0) + D = 0 defines a plane. A, B, C are normal.

Orthogonality property: two vectors (V1 and V2) are orthogonal if their dot product is zero.

V1.V2 cos (theta) = 0. This equation is also used in Dynamics section but with respect to mass.

Intersection of lines[9] external source

See footnote 9.

---

9. Rogers and Adams, *Mathematical Elements in Computer Graphics*, 69.

Dot product and cross product

Dot product is be used in section 5 on iterative methods. Dot product with two vectors **a** and **b** is given by:

$$a.b/|a||b| = \cos(\text{theta})$$ Equation 4

Theta is the angle between two vectors.

Cross product between two vectors 'a' and 'b' is given by:

$$\mathbf{axb} = \begin{vmatrix} i & j & k \\ a1 & a2 & a3 \\ b2 & b2 & b3 \end{vmatrix}$$ Equation 5

$$i = a2*b3 - a3*b2$$
$$j = a1*b3 - b1*a3$$
$$\mathbf{k = a1*b2 - a2*b1}$$

Figure 23. Cross product.

Construction of Model

Construction of graphic model typically uses geometric features listed table 2 below. Using the line feature and maintaining spatial distances between the objects, the algorithm develops a prism from the geometric parameters. It subsequently merges these prisms into making a full model (figures 20 and 31).

Hidden face removal[10]
(Reader may want to refer to the footnote)

Graphics geometry: for list of geometric entities see table 2.

---

10. Angell and Griffith, High-Resolution Graphics in FORTRAN 77, 226.

Model geometry: A CAD geometry for this model consists of multiple lines as geometric feature. Alternatively automatic mesh generation is used and modified for making new objects.

Table 2. CAD elements.

Geometry Type	Description
Point	Is an entity in space with an x, y, z coordinates
Line	Typically, is a connection between points
Normal	Unit normal
Curves	Is a geometric entity associated with a curvature attribute
Cubic spline	Representation of curve with a third-degree polynomial.
B-Spline	Representation of curve with Control Points and Knot vectors
Knot vector types	Uniform knot vectors, open knot vectors, non-uniform knot vectors
Rational B-Spline	Ratio of polynomials
Conic Sections	Intersection of a plane with Cone Makes circles, ellipse, parabola, hyperbola
Surfaces	These are an Orthogonal net of curves
Plane	Equation of plane
B-Rep-Solid	Is a package comprising of all the faces of solid object

Algorithm for Curves

The following types of curves are basic to geometry:

Cubic spline
B-spline
Rational b-spline

Algorithm cubic spline[11]

Cubic spline curves are formed from the following basis function:

$$P(t) = a0 + a1\ t + a2\ t^2 + a3\ t^3 \hspace{3cm} \text{Equation 6}$$

In the above equation "t" is a parameter and a0 to a3 are constants; thus, this is a combination from four types of entities. Boundary conditions are imposed at two ends of the curve to determine the constants and to get a continuous curve. Practically, beam analysis uses a cubic polynomial as a spline trial function for determining displacements. FEA on beam made up of multiple line segments, uses a similar approach, where the boundary conditions are typically slope and displacements. Through a sequence of transformation, the constants a0 to a3 are transformed to unknown displacements at nodes. Equilibrium between internal and external disturbance results in an equation that is further solved for unknown values at the nodes by a process of minimization and leading to an evaluation of constants. This is described in the Solid Mechanics section with basics of FEA equations.

Displacement curve is made up of a rigid body translation, a straight line, a quadratic curve, and a cubic curve.

P (t) is a point vector on the curve.

Bezier curves

See footnote 12, page 289 in the reference.

Algorithm for B-spline curves[12]

Computer program for B-spline is enclosed in graphics software (in concept). Excerpts are requoted here. We present a graphics form, to this algorithm referenced above.

(from footnote 12 below)

> "The formulation of a B-spline curve is based on **B-Spline basis function** developed from knot vectors and control points. The degree and order of the curve is an important element. Order is the same as number of control points.

---

11. Rogers and Adams, *Mathematical Elements in Computer Graphics*, 253, 289.

12. Rogers and Adams, 305–316.

> The basis values are generated recursively from knot
> vectors from previously determined basis functions of
> each level... Basis functions are developed between 2
> support points.
>
> The knot vectors are classified as **open knot vectors,
> uniform knot vectors** and **non-uniform knot vectors.**
>
> The degree depends upon the control points. Degree
> *t is always one less than the control points.* knot vectors
> are developed based upon the number of basis functions
> desired. [13]"

An example below demonstrates the recursive basis function from knot values. This algorithm by Cox-deBoor is a recursive algorithm.

A point on the curve P(t) is defined as:

P (t) = B (i,) * Sum (Ni, k) [t]

Where: B [I] = control points

Ni (t) = basis functions, t being a point between intervals
With the stipulation that Ni, 1 = 1 when I is the first interval in
the segment and Ni,1 = 0 when I is not within this limit.
Thus Ni,1 is a starting point.
The system can be built from the following:

N1,1 =1
N1,2 from N1,1
N1,3 from N1,1 and N1,2
(For further logic, see reference cited above.)

---

13. Rogers and Adams. 306

Cox-deBoor recursive formula:
(from footnote 13)

$$N(I, k)(t) = [(t- x(m)] * N(I,k-1) (t) / [x(m+k-1) – x(m)]$$

Equation 7

$$+ [x(m+ k) – t]*N(I+1,k-1) (t)/[x(m +k) – x(m+1)]$$

Where: N(I, k) is the next basis.
m is next knot value from knot vector x(m)
k, index to next degree
Numerator is based on the product of basis function and parameter (t).
Thus, the next basis function is developed from the previous value.

A graphical representation is shown below to clarify the statements above.
The value at a point (t) on a planar curve or spatial curve is result from all the basis curves and control point, and is expressed as:

$$P(t) = Sum [B(i)*N(I, k)]*t$$

For each "t" between initial (0) and max value. For this example t(0) to t(max) is subdivided into six segments controlled by required basis function for a third degree curve.

Typically, expanding the above expression for a control point of four leads to an expression:

$$P(t) = B1\{N1,k(t)\}+ B2\{N2,k(t) + B3\{N3,k(t) + B4\{N4,k(t)\}$$

Equation 8

Where 't' is a parametric point within the first control point and last control point.

B-spline algorithm with seven knot vector elements.[14] Recursive procedure follows forward and backward values of the tree (figure 24 to 29) and is based upon the control points and upon Cox-deBoor equation. Knot vectors form a dependency for forming basis functions; for example, if the control point are four points, then the number of knot vectors required will be 4+3=7 to satisfy the Cox-deBoor equation.[15]

The diagram below is for a uniform knot vector [0, 1, 2, 3, 4, 5, 6]. It has six intervals: (0 to 1), (1 to 2), (2 to 3), (3 to 4), (4 to 5), and (5 to 6). The basis function is N(i, j), for each interval uses Cox-deBoor equation where "j" max is 3 for third-degree curve.

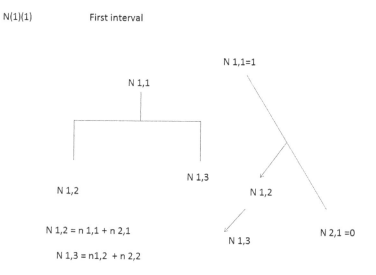

Figure 24 . First level computation of basis function N(i, j).

---

14. Rogers and Adams, 315.

15. Rogers and Adams, 306

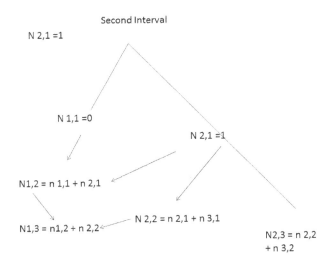

Figure 25. Second level computation of basis function N(i, j).

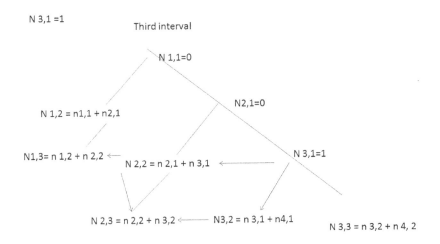

Figure 26. Third level computation.

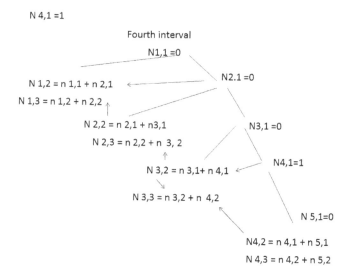

Figure 27. Fourth level computation.

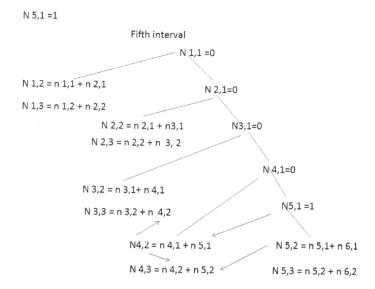

Figure 28. Fifth level computation.

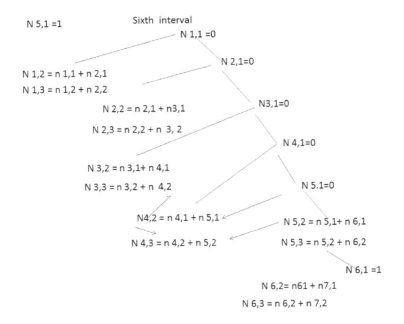

Figure 29. Sixth level computation

Algorithm for Rational B-Splines[16]

Rational B-splines have another parameter "h" that allows control over local spline curve.

B-Splines (NURBS)

This is a reduction from a four-dimensional space to a three-dimensional space. The spatial coordinates are x, y, z, and h; thus, the numerator is one dimension higher than the denominator.

We start building the curves from a B-spline formulation. In building these curves we make use of knot vectors. A knot vector is a row of real numbers increasing in numbers. The purpose of knot vectors is to allow

---

16. Rogers and Adams, 356

the curve to pass through the points on the knot vector. Knot vectors are defined as uniform, open, and non-uniform vectors.

Cox-deBoor formula is applied to develop basis function:

$N(i, k)*t = (t- x(m) * N(m,k-1) *t / x(m+k-1) – x(m) + [x(m+k) – t]*N(m+1,k-1) * t / x(m+k) – x(m+1)$

In the case of rational B-splines, we have an additional dimension called as "h"; thus, we make four-dimensional space by choosing a matrix of four rows and four columns. The last column being the value of "h," which allows control of curve in a localized fashion.

In the case of rational B-splines, each control point is tied with a vector [h]. The numerator is defined with B[i] h[i] N[I, k] where [B] is a list of control points and [h] is a factor associated with control point to change the nature of curve between selected control points. The denominator is defined with [h]N[I,k]; thus, a point P(t) for the curve is defined as:

$P(t) = B[I] h[i] N[I, k] (t)/h [I]N[I, k]$ Equation 9

Where I is the set of control points.

All the basis functions defined previously for B-spline are still valid but are modified with h[I] vector associated with each control point.

Surface

Algorithm for Surfaces[17] external source

The equation of a quadratic surface is given below:

$Ax^2 + By^2+Cz^2 +D x +E y+ F z +G+H+I =0$ Equation 10

The nonexplicit version of a surface is defined as **sweep surface**, which is created by sweeping a line in space. Curves when swept can generate a surface.

---

17. Ibid., 379, 400.

For constructing a surface from B-spline in a two-dimensional space, see footnote 17. B-spline surface is created by applying Cox-deBoor algorithm in two dimensions and typically on two planes normal to each other.

Graphical examples of surfaces

Example of a planar surface in x y plane is shown in **figure 30.**

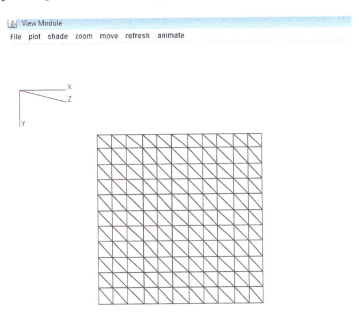

Figure 30. Planar surface.

Algorithm for Three dimensional Clipping[18] external source

(Reader may want to refer to the cited reference)

Data structure for the graphics' module is composed of following:

File name: **object.dat**->data structure

---

18. Angell and Griffith, *High-Resolution Computer Graphics Using FORTRAN 236*

Number of objects
Object definition
Number of vertices of the object
Initial origin of object x, y, z
x, y, z of each node with initial displacement
Total number of faces
Total node of each face
Define face with the node

Attached is a sample of Object.dat file below (full file is saved on thumb drive labeled as "Assembly.java"):

72 (number of objects)
Left vertical support object 30: 9 (object definition)
8 1 (number of vertices of the object; 1 is used for controlling rotations)
21.0 10.0 1.0 (initial origin of object x, y, z)
1 0.0 0.0 0.0 (local x, y, z coordinate of each node)
2 0.0 12.0 0.0
3 0.0 12.0 6.0
4 0.0 0.0 6.0
5 1.0 0.0 0.0
6 1.0 12.0 0.0
7 1.0 12.0 6.0
8 1.0 0.0 6.0
6 (total number of faces of each object)
4 (total node of each face)
1 4 3 2 (define face with the node)
4
8 7 6 5
4
5 8 4 1
4
6 7 3 2
4
5 6 2 1

4

8 7 3 4

2nd vertical diagonal 20; 31

12 0

31.0 20.0 2.5

1 0.0 0.0 0.0

2 0.5 0.0 0.0

3 2.5 2.0 0.0

4 2.5 2.5 0.0

5 2.0 2.5 0.0

6 0.0 0.5 0.0

7 0.0 0.0 0.5

8 0.5 0.0 0.5

9 2.5 2.0 0.5

10 2.5 2.5 0.5

11 2.0 2.5 0.5

12 0.0 0.5 0.5

8

4

6 5 11 12

4

1 6 12 7

6

1 2 3 4 5 6

6

7 8 9 10 1112

4

1 2 8 7

4

2 3 9 8

4

3 4 10 9

4

4 5 11 10

Top bar angle 1:12

8 0

26.0 20.0 2.5

1 0.0 0.0 0.0

2 12.0 0.0 0.0

3 0.0 0.0 0.5

4 12.0 0.0 0.5

5 0.0 0.5 0.0

6 12.0 0.5 0.0

7 0.0 0.5 0.5

8 12.0 0.5 0.5

6

4

4 3 1 2

4

8 6 5 7

4

7 8 4 3

4

5 6 2 1

4

5 7 3 1

4

8 4 2 6

Vertical side of top bar 1: arm 2; 13

8 0

26.0 20.5 2.5

1 0.0 0.0 0.0

2 12.0 0.0 0.0

3 12.0 0.0 0.25

4 0.0 0.0 0.25

5 0.0 0.5 0.0

6 12.0 0.5 0.0

7 12.0 0.5 0.25

8 0.0 0.5 0.25

6

4

1 2 3 4

4

5 6 7 8

4

1 2 6 5

4

4 3 7 8

4

1 4 8 5

4

2 3 7 6

Applications

Geometric example is added and subsequently crafted. A graphic user interface screen is displayed in figure 31. Left side shows all the objects of the model; the center of screen is a display showing options to activate.

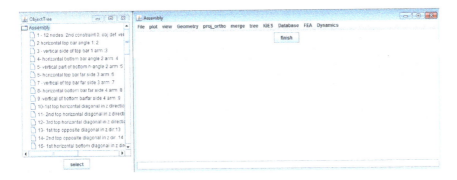

Figure 31. Typical image from collection of objects of the crane model on left.

Right side of the above figure displays options within integrated environment.

Road Map of the Program

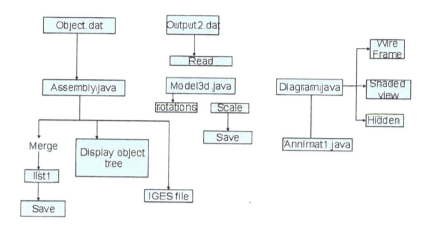

Figure 32. A snapshot displaying a process for graphics model.

Steps for building a graphics model:

1. Object.dat is the first file for geometry and is read in Assembly.
2. Assembly makes a new file and saves as "output2.dat" and "saveOb-jectToArc.dat" (for Annimate.java)
3. Model3d reads output2.dat file and applies viewing parameters and makes product_file.dat file that holds the faces of the model.
4. Diagram reads product_file.dat and displays the model.
5. Animate will read from readFEAfile.dat while in paint graphics method. This function couples FEA with animation of model arm.

File structure of product_file.dat used for displaying model is shown below[19]:

8 nodes

24 arcs

1 128.0 291.0 (two-dimensional node coordinates)

2 9.0 392.0

---

19. Data structure for the file is adopted from Keith Weiskamp, Loren Heiny, and Namir Clement Shammas, *Power Graphics Using Turbo C* (New York: John Wiley and Sons Inc., 1989). W/permission on page 340.

3 460.0 392.0

4 341.0 291.0

5 128.0 110.0

6 9.0 9.0

7 460.0 9.0

8 341.0 110.0

1 4 3 -2 (arcs)

6 7 8 -5 ("- ve" circles back to the first node)

5 8 4 -1

3 7 6 -2

8 7 3 -4

1 2 6 -5

Example Using Rectangular Finite Element Model

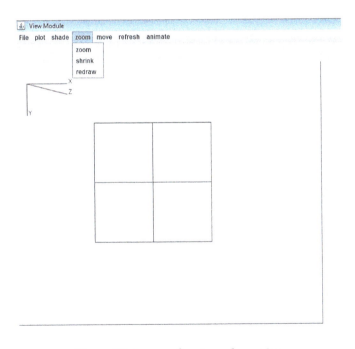

Figure 33. Image of rectangular region.

Data Structure for the Above Model

Using mesh generation, an automatic mesh is generated from Java program. This file is the same as output2.dat but referenced with nodedataR.dat to be more specific for rectangular element and for input into **Model3d.java** where other functions are carried out for display.

Typical Structure of File

NodedataR.dat saves the number of nodes, its coordinates, and "face to arcs" relation between nodes:

9 (nodes)
1 0.0 0.0 4 1 (coordinates node, x, y, z, of object)
2 3.0 0.0 4 1
3 6.0 0.0 4 1
4 0.0 3.0 4 1
5 3.0 3.0 4 1
6 6.0 3.0 4 1
7 0.0 6.0 4 1
8 3.0 6.0 4 1
9 6.0 6.0 4 1
1 2 5 -4
2 3 6 -5
4 5 8 -7
5 6 9 -8

The subsequent file name is called product_file.dat after executing Model3d and saving the file for viewing the model in Diagram Java. (Input file is converted to 2D file.) Data structure for product_file.dat:
9 (number of nodes)
16 (number of connected arcs)
1 12.0 -9.0 1 (node coordinates x, y, z in PC view coordinates)
2 49.0 -9.0 1
3 87.0 -9.0 1
4 12.0 27.0 1
5 49.0 27.0 1

6 87.0 27.0 1
7 12.0 65.0 1
8 49.0 65.0 1
9 87.0 65.0 1
1 2 5 -4
2 3 6 -5
4 5 8 -7
5 6 9 -8

Space Truss Data for Specific Angle

Case 1 for Verification of Geometry

Data is generated from animation with the file name:

nodedataR_sptrus_for_angle11.885692533.dat

This data for truss is generated from Annimat1.java; the data is edited. Data is for an angle **angle11.88** (figure 34).

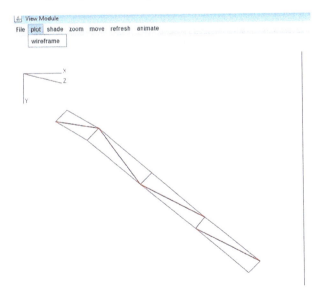

Figure 34. Image verifying geometry.

Observation

From the image (Figure 34), it is verified that the non-diagonal member meets at right angles to chords and maintains the geometry from animations. This file is exported to Fortran program for stress analysis. More on analysis is covered in the section called Solid Mechanics.

We built truss data using the rectangular element outline; we added additional connection using this basic outline from automatic generated model. Any addition of arc requires four nodes.

Typical data structure of file exported for stress analysis:

Data>

10 (total nodes)

1 0.0 0.0 4 1 (typical node coordinates)

2 23.0 18.0 4 1

3 46.0 37.0 4 1

4 69.0 56.0 4 1

5 93.0 75.0 4 1

6 4.0 5.0 4 1

7 18.0 23.0 4 1

8 41.0 42.0 4 1

9 65.0 61.0 4 1

10 88.0 80.0 4 1

1 6 7 -2

2 7 8 -3

3 8 9 -4

4 9 10 -5

Modified data (bold items have been modified):

11 (increase in number of nodes)

1 9.0 10.0 4 1

2 23.0 18.0 4 1

3 46.0 37.0 4 1

4 69.0 56.0 4 1

5 93.0 75.0 4 1

6 4.0 15.0 4 1

7 18.0 23.0 4 1

8 41.0 42.0 4 1

9 65.0 61.0 4 1

10 88.0 80.0 4 1

11 0.0 0.0 4 1

1 6 7 -2

2 7 8 -3

3 8 9 -4

4 9 10 -5

6 2 6 -2 (diagonals)

5 9 5 -9

2 8 2 -8

8 4 8 -4

11 5 11 -5

(For the inclined rope, see figure 38 in Graphical Images.)

Case 2 Plane Truss

Figure 35 shows another test model for plane truss generated from automatic rectangular element and modified for diagonals.

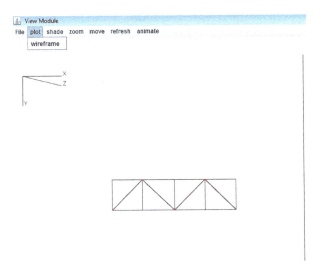

Figure 35. Image of planar truss.

The stress analysis for this data is in the Solid Mechanics section and thus not reproduced here.

Dynamics of Graphics Model

Procedure

The following operations are carried out on **Assembly.java:**

1. File is imported with the definition of all the objects and the parameters.
2. Merge function merges all the objects defined in the file (**from a predefined file**).
3. In table 3, merge function in **Assembly** saves the file to output2.dat file. **Model3d** reads this file, applies rotations and scaling, and saves it to product file.dat, which is read in **Diagram.java** for viewing the graphics.

Table 3. Program actions.

Description of Java Files	Description of Functions
Assembly	**Object merging**, and save
Model3d	Change view angle, apply scaling and save
Diagram	View model and apply rotations, adjust parameters for perspective view
Annimat 1	Run Animation

Figure 36. Viewing options by rotations.

Figure 37. Tool for changing values.

The graphics model displays a three-dimensional view of the object; thus, one can specify a parameter for rotation of about three axes. The graphical tool shown gives a means of adjusting these parameters. These are rotations about three axes. Thus, the observer can view the model from different angles, in perspective mode, or look at larger view by zooming. This is achieved by mathematically manipulating the objects with transformations. (see Table 4)

Table 4. Functions.

Available Functions	Actions	Method
On Diagram.java file	Multiple viewing with motion button	
Multiple Rotation	About x, y, z axis	**From cited references**
Zooming	With a slider	
Perspective viewing		
Scaling	With a slider	

Graphical Images

Side view

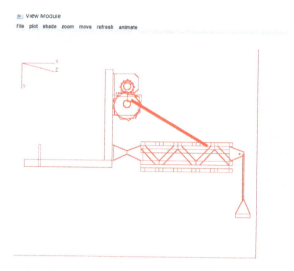

Figure 38. Wire frame view of model displayed in one plane (X, Y).

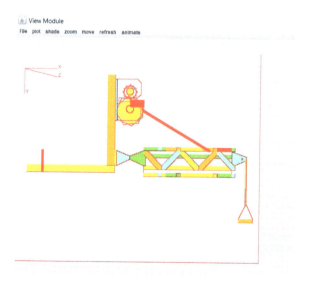

Figure 39. Shaded view of model displayed in one plane (X, Y).

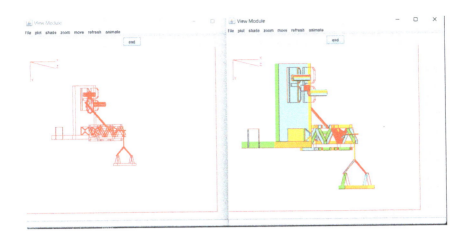

Figure 40. Image of model in wireframe and shaded view.

Parameters for perspective views[20]

"Perspective view uses manipulation with three 4x4 matrix. Two of the matrices are for rotation First about X and then about Y axis. A third matrix 4x4 matrix is used for a single point parameter for perspective projection." Interested reader may visit Roger's book cited in footnote 20.

"PersP" function in "Model3D.java" sets up the data.

Two functions place the image in View panel. One is scaling and saving in "Model3D.java".

Second is viewing the image , which is done in "Diagram.java" using "File open " and "Plot" function.

Image may not appear on the view panel. The function "center" is used to bring the image to the "View Panel", "zoom" and "shrink" tools are used to resize the image.

This function is activated on "Diagram.java" with "move" button. Images are distorted in perspective views.

---

20. David Rogers and Adams pages 159, 177
    Keith Weiskamp, Loren Heiny, and Namir Clement Shammas, *Power Graphics Using Turbo C* (New York: John Wiley and Sons Inc., 1989) with permission on pages 341,348,349, 350,351,352,354

Alternative method for perspective projection is described in Weiskamp's book. (See footnote on second author).

These views confirm the outline of the model.

Interested reader may refer to these pages for more information.

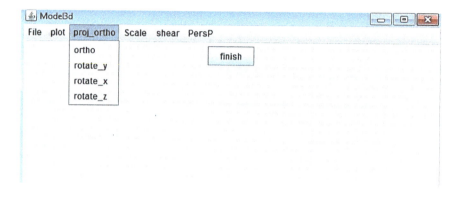

Figure 41. Typical options for rendering in perspective view.

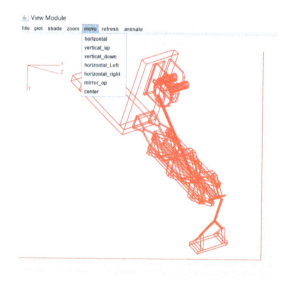

Figure 42. Image with positioning options.

Perspective view

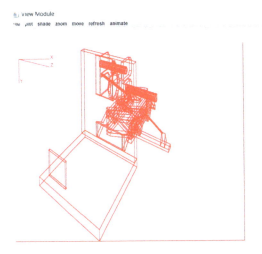

Figure 43. Perspective view of partial model in wire frame.

Shaded perspective view

Figure 44. Partial perspective model with shading.

Illumination and shading

We adopt an algorithm of point "inside and out", on a bounded region for shading[21]

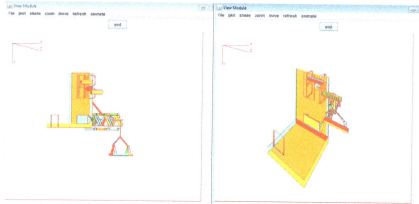

Figure 45. Composite images in orthographic projection and perspective view.

21. Angell and Griffith, *High-Resolution Graphics Using FORTRAN* 50, 129 w/ permission

Animation

Animation allows the study of motion of the model and establishes clearances between the parts while in motion. It allows testing of several parameters that need to be observed, typically stresses in various members. To run, the graphics programs they are listed in table 5.

Table 5. Batch file names at command prompt.

Function name	Argument (Batch file name)	Result
d>fnprj (selected path)		
d>fnpri	Cadcam	For main Screen
d>fnpri	Or Assembly	For main Screen

Finite element analysis

We capture the state of running boom into files and perform a finite element analysis as a run-time example. This can be viewed on Finite Element section. Kinematics and kinetics are coupled; thus, animation is a combination of two.

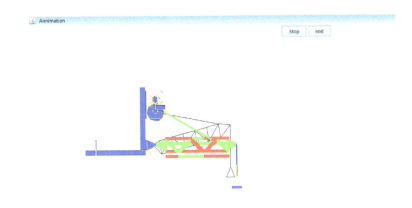

Figure 46. Typical image from animation.

Animation and stress analysis

The image for particular orientation in x, y plane is in figure 46 above.

We select 10 instances for truss analysis with animations. The next segment for FEA is integrated with graphic movement of crane arm and simultaneous stress analysis. This demonstrates a run-time effect of stress analysis on the model (see views in part 4: Solid Mechanics).

Automation: RF Operations

The physical model is operated with RF control. For more details, user can visit the website on the thumb drive, enclosed in HTML.

Conclusion

We created an outline of a model and performed various projections for viewing the model. We added animation and shading, eventually linking animation with FEA. Finally, we fabricated a model to the outline. We added basics of computer graphics, such as splines and commonly used equation in computer graphics (planes). We referenced other text as a source for further reading.

PART 4:
Solid Mechanics

Numerical Analysis of Objects: Framework[22]

In this section, an example is presented where the model arm is dynamically simulated at run time for stresses, coupling with animation.

Background

Finite Elements (CAE) and Experimental Stress Analysis

A dynamic model is created, coupling graphic animation and FEA and adding experimental stress analysis at postfabrication. For more on this, see section 7 on experimental stress analysis.

---

22. See: K. C. Rockey, et al., *The Finite Element Method: A Basic Introduction for Engineers* (New York: John Wiley & Sons, 1983); Weaver Jr., William and James M. Gere. *Analysis of Framed Structures*. New York: Van Nostrand Reinhold Co., 1965.Pin Tong and John Rossettos, *Finite Element Method: Basic Technique and Implementation* (New York: Dover Publications, 1977); J. N. Reddy, *Introduction to the Finite Element Method* (New York; McGraw-Hill, 1984); and O. C. Zienkiewicz, R. L. Taylor, and J. Z. Zhu, *The Finite Element Method: Its Basis and Fundamentals* (New York: McGraw-Hill, 1967).

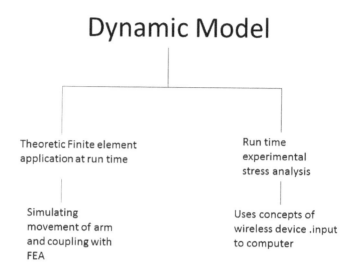

Dynamic Model

Theoretic Finite element application at run time

Run time experimental stress analysis

Simulating movement of arm and coupling with FEA

Uses concepts of wireless device .input to computer

An introduction to finite elements methodology

The finite elements methodology is another vast area. A finite element is a decomposition of a region of solid or fluid body into smaller regions. These smaller regions are called finite elements. This applies to both a stick element as well as a continuous body.

The purpose of breaking a region into smaller regions is to mathematically find a solution to the problem of a small region and then extend it to an overall area of interest by summation. Such an area is called domain of the problem. Eventually, use of these results in structural mechanics to analyze the displacements, principal stress, and shear stress. These quantities are used to indicate the sustainability of the materials to external disturbances. The theoretical development of FEA is based on Rayleigh-Ritz method[23] or Galerkin method of performing finite element analysis. The analysis renders itself to applications with cut-outs and irregular boundaries that do not cause singularity.

---

23. Chi-The Wang, *Applied Elasticity* (New York: McGraw-Hill, 1953).

Combining theoretical numerical analysis with experimental phase pertaining to the model is the distinguishing feature of this project. This exhibit is model centric. Theoretical analysis is a result of analysis from the files created from graphics model during animation.

The experimental phase will have to interface with LabVIEW and wireless acquisition of data initially acquired from a beam model. It is further projected to the CAM model and real-world application (in this case, using a quarter-bridge strain gauge configuration of Wheatstone Bridge

Mathematics of finite elements are not reproduced here; however, salient features that go to make finite elements are addressed. These features include potential energy function, method of solution, interpolation functions, and direct displacements method. The method of solution adopted here is with Gaussian elimination on a semi bandwidth.[24] The writer supplemented this method with his program on Gaussian elimination method (Gsolve2.for) for solution on a fully populated symmetric stiffness matrix.

Compatibility can be viewed by comparison of theoretical results and experimental stress analysis.

The application of finite elements has spread across many engineering disciplines. These being structural, electrical, and fluid mechanics, and heat transfer. In essence, any system that can be described by a second- or fourth-order ordinary, or second- or fourth-order partial differential equation with boundary condition, is a candidate for finite elements, including nonlinear elastic properties. The method is general such that numerous ordinary and partial differential equations can be subjected to this method for solution.

In this finite element analysis, we perform an analysis on a motion of the boom. For each position of the boom, we do a stress analysis. Our objective is to verify these values with the experimental phase described later in the sections but performed on a beam model. The total energy of the system is shown in figure 53. We identify the elements on the model shown in figure 51.

The complete finite element model will be a combination of unknowns at the node—that is, displacement or degree of freedom matching at the nodes. The equation to solve will be the total energy of the system. We

24. W. M. Jenkins, *Matrix and Digital Computer Methods in Structural Analysis* (New York: McGraw-Hill, 1969). 171-173

solve each individual component by running the appropriate program (as shown on graphical integration page, figure 59). For simplicity, we analyze the moving arm as space truss only. For interested readers, the continuum part can be added by adding an element type to the software (appendix E shows this option). Individual programs are presented on the integration screen display.

We identify various finite element components in the model (figure 51). Subsequently we develop the finite element of each of the components (in the software).

We then send pointers to the finite element mathematics (in cited references of appendix B) and draw its application on those components identified on the model. The stiffness matrix that is applicable to each element thus identified are referenced in appendix B.

The figures below display typical finite element mesh for plane stress and plate bending problems with rectangular and triangular mesh on a continuum.

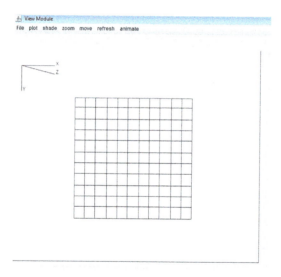

Figure 47. Typical subdivision of a region with rectangular elements.

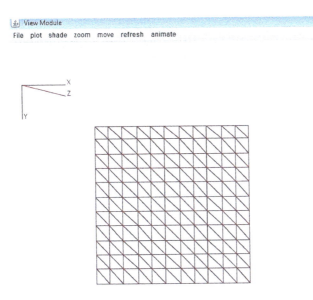

Figure 48. Typical subdivision of a region with triangular elements.

The above images are typically associated with problems regarding plane/stress plane strain and Kirchhoff's plate bending.

A general formulation of FEA

The algorithm shown below is an operator equation:

$$T u = f \qquad \text{Equation 11}$$

Where: T is an operator

T can be a second- or fourth-order differential operator.

u is an assumed trial function in terms of polynomial cited in the polynomial section.

f is the external disturbance.

Selected polynomial will depend upon the problem to address.

Central to this is to convert the operator T into a stiffness matrix. This has several approaches from using direct displacement, the potential energy method, the virtual displacement method, and complimentary energy methods.[25] For example:

u = a0 + a1 x + a2y

The above equation is a polynomial for a triangular element, with "u" taking the place of an unknown for one degree of freedom. Transforming the T (del) operator in terms of transpose and the column vector by taking the partial derivatives or ordinary derivatives of the equation is the next step. This is followed with integration because both x and y vary over the given region of individual elements.

The next step is followed with a summation process from individual elements to form the global stiffness matrix for the entire region of interest. Mathematics for this derivation can be found in noted books on introduction of finite elements listed in the bibliography.

This is a lengthy process leading up to building a stiffness matrix. Only key equations applicable to model are retained in this manual.

Next would be to apply a minimization process to establish equilibrium between internal and external forces, and apply appropriate boundary conditions to solve the unknown such as "u" or displacements (see figure 54). Typically, this construction results in a symmetric matrix as shown below (figure 58). In general, this equation can be rearranged as the equation below (Eq. 12).

In elasticity problems, G is derived from strain displacement and stress-strain relationship, using Hooke's law of elasticity. This is rearranged as Gt and Gc, as transpose and column matrix for solutions of the unknowns. This process is lengthy, thus, only the resulting equation is presented (see footnote 27). Typical evaluation on a fourth-order differential equation results in an expression below (equation 12).

Applying the method of virtual work or minimization method of the differential integral form results in a structure:

---

25. For more information, see K. C. Rockey et al., *The Finite Element Method: A Basic Introduction for Engineers;* Pin Tong and John Rossettos, *Finite Element Method: Basic Technique and Implementation;* and William Weaver Jr. and James M. Gere, *Analysis of Framed Structures.*

$$\int \{u\}t \, [Gt] \, [P][\, Gc \,]\{u\} = [F]\{u\} \qquad \text{Equation 12}$$

Where "t" is used to indicate a transpose.
Operator (T) is arranged as Gt, a transpose, and Gc as column vector of the operator equation(11). This same method also applies to second-order ordinary or partial differential equations.

In general, P is the property matrix. P matrix is a property matrix and for two-dimensional problems can be expressed as:

$$\begin{bmatrix} P11 & P12 \\ P21 & P22 \end{bmatrix}$$

Figure 49. Property matrix.

[P] is the problem-specific property. In problems of elasticity [P] can relate to elasticity or elasticity and poisons ratio defining plate rigidity, in heat transfer problems [P] can be the thermal conductivity, in flow problems through porous media it can take up the soil's permeability matrix,

This expression (equation. 12) when integrated over the volume results in the stiffness matrix of the element. This is defined as K—stiffness matrix—and is symmetric in most cases, resulting in an equation.

$$[K]\{u\} = \{F\} \qquad \text{Equation 13}$$

This equation (equation 13) is solved for unknown nodal values using proper boundary values. Boundary values depend on the physics of the problem.

We demonstrate the application of this equation on a physical model. A similar pattern (i.e., equations 12 and 13) also result for the ordinary fourth-order (beam) and partial differential equation, such as in plane stress and plane strain and plate bending problems. Followed in this segment are finite elements for a run-time simulation of a component of the model (Coupling with Graphics animation page, figures 46 , 70 to 72).

Other applications

There are several problems known as field problems that are defined by second-order partial differential equations that can be solved by this method. They are summarized below:

Torsion of a bar
Fluid flow problems
Poisson equation
Prandtl equation
Darcy's flow in soil
Heat transfer

Next we capture an image of the boom from animation.

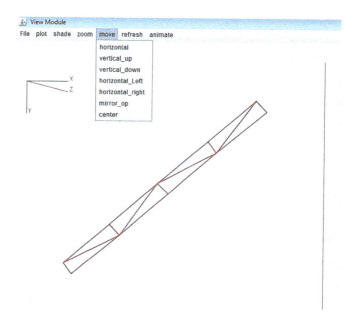

Figure 50. Typical analysis on a component of the model, modeled as an assembly of truss element for runtime simulation.

Formulation of differential equation for beam. Before we apply the method, we need the system differential equation. For example, we use Euler-Lagrange for beam and the Kirchhoff plate-bending equation as well as biharmonic equation for two dimensional problems. For string this would be second order differential equation. Our approach uses a polynomial in describing the unknown value.

Alternatively, interpolation methods can be used to describe the generalized coordinates. In this approach, the unknown take on the unit value at each node and vary over the element via interpolation function.

True solution to the problem satisfies compatibility and meets the boundary condition Compatibility in general is not a problem with Euler-Lagrange equation; however, it is generally a problem in two-dimensional regions, specifically in plate-bending problems. These are treated in more advanced finite element methods[26] and referenced in the bibliography. Applicable methods of finite element are summarized below.

Summarizing other methods

We focus on the theory of displacement analysis.[27] The following is a list of alternate numerical methods:

Finite difference method
Complementary energy method (analyzed by assuming stresses as unknowns)
Virtual work method
Direct displacement method
Rayleigh-Ritz potential energy method (assumes displacements as unknowns)
Least squares method
Galerkin method
Weighted residual method
Reissener method (unknowns as stresses)
Mixed method

26. Tong and Rossettos, *Finite Element Method*. 230-231
27. See William Weaver Jr. and James M. Gere, *Analysis of Framed Structures*, Tong and Rossettos, *Finite Element Method*;; K. C. Rockey et al., *The Finite Element Method*; O. C. Zienkiewicz, R. L. Taylor, and J. Z. Zhu, *The Finite Element Method: Its Basis and Fundamentals*; and J. N. Reddy, *Introduction to the Finite Element Method*.

Typical applications of finite elements have been addressed in the following areas:

- Design of structures
- Eigenvalue problems (seismic analysis of structures with FE)
- Run-time simulation with FE
- Finite element and CAM
- Folded plates by FE
- Plate on elastic foundation
- Models with Penetrations or cut out from a body as finite element.

Pointers are given in appendix B on some of these, as it would be beyond the scope to reproduce them in this manual. We essentially use method based on **potential energy** and direct displacements.

Element types associated with the model are identified in the figure 51 below.

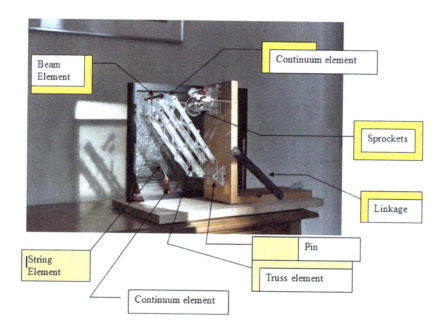

Figure 51. Showing various types of element.

For the model shown in figure 51, the total energy is formed from the contribution of individual element. Their corresponding images are shown in figures 53 and 54.

A polynomial approach is taken for the development on continuum. The unknowns are described in terms of approximate function made up of power series in Cartesian coordinates or in polar coordinates, as the case may be.

Typical polynomials are listed below.

Polynomials used in FEA for Displacement method

Axial displacement = a0 + a1 X

Beam Element = a0 + a1X + a2 X^2 + a3 X^3

Plane Stress : u = a0 + a1 X + a2 y

v = a4 + a5 y + a6 X

Plate Bending = unknown rotations and displacements

[12x12] matrix for rectangular element

[9x9] for triangular element

3D -Brick Element =[24x24] matrix of displacements

3D -Tetrahedron = [15x15] matrix of displacement

Figure 52. Typical polynomials in FEA.

a0 to a6 are typically constants associated with plane stress/strain problems. These constants associated with the polynomials are determined by matrix inversion(Appendix B, typically shown on developing a beam stiffness matrix). Subsequently strain displacement relation is used in leading up to an evaluation of unknown nodal values. Nodal values are determined from a series of transformation, typically using strain-displacement and stress-strain relationship and establishing an equilibrium equation by minimization[28], as shown in figure 54.

---

28. Chandrakant S. Desai, and John F. Abel, *Introduction to Finite Element Method: A Numerical Method for Engineering Analysis* (New York: Van Nostrand Reinhold Co., 1971).,pages 8-9

Nodal values are found by solving the matrix equation (FEA, equation 13).

Total Energy of the System

Stiffness of then truss elements   +   Stiffness of the continuum elements

$$[K_{system}] = [K_{truss}] + [K_{continuum}]$$

Figure 53. Analytical model.

Equilibrium Equation

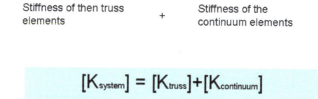

Π =   [PI], Potential Energy function

Π =   (½) $[K]^*\{u^2\}$ -   {P} u

$\delta\Pi$ = $[K]\{u\}\frac{\partial}{\partial}u - \{P\}\frac{\partial}{\partial}u^\mu$ =   0

                          $\frac{\partial}{\partial}u \neq 0$

$[K]\{u\} = \{P\}$

Figure 54. Equilibrium equation.

FEA Fundamentals

The equation we wish to solve for Static Analysis

$$[K]\{u\} = \{F\}$$

Figure 55. FEA equation.

Definitions

Where
[K]= the stiffness matrix
{U} = displacement vector
{F} = force vector

Figure 56. FEA definitions.

The analytical model is solved by applying appropriate boundary conditions, such as fixity, hinged, roller, or other (i.e., by introducing spring constants).

[K] is a stiffness matrix and shown in the figure 58 below. It is a summation process of degrees of freedom from the individual contributing element (figure 57) and assembled into a global stiffness matrix for solution. (Figure 58)

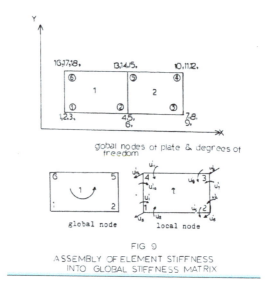

FIG 9
ASSEMBLY OF ELEMENT STIFFNESS
INTO GLOBAL STIFFNESS MATRIX

Figure 57. Figure identifies degrees of freedom typically on two plate elements.

[K] is a global stiffness matrix, is a sparse matrix, and shown in figure 58 below.

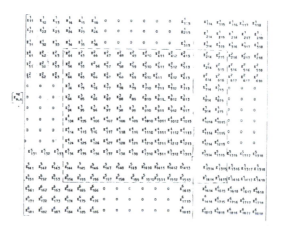

Figure 58. Global stiffness matrix a summation process from degrees of freedom.

Table 6 below shows some of the elements in this manual. Our finite element model is an animated finite element. It develops the finite element analysis for various positions of the moving arm. The approach that we take is one of displacement method and minimizing the potential energy.

The model showed in figure 51 uses truss and continuum elements for analysis. The analytical model developed from figure 51 is made up of truss elements that are used for simulation. Figure 59 shows an integration of the FE models with Java GUI. Here, the selected finite element program is executed from Java graphical user interface screen.

The figure below is from a screen capture, displaying options.

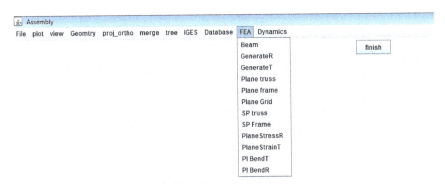

Figure 59. Fortran programs on FE integrated into Java.

Table 6 lists typical finite element types.

Table 6. Finite element types.

Element Type	Purpose	Degrees of freedom
Discrete Stick element	**Applications to Truss - 1 degree of freedom**	1
Beam element	**Applications to beam -2 degree of freedom**	2
Space Truss	**Application on space truss**	3

Space Frame	**Application on Frames**	6
Plane Stress/Plain Strain Element	**Application- on in plane problems with 2 degree 2 of freedom**	2
Plate Bending	**Application to continuous body with normal pressure, such as foundations resting on piles, floor slabs**	3
Folded Plate	**shells**	6
Plate on elastic foundation	**From plate bending**	3
Related topics of vibration		
Eigen value problems		

In this manual, examples include the subtopics listed in the table 7.

Table 7. FEA models and proofs.

Proof of beam model	Shown in theoretical and Experimental stress analysis
Proof of Plane truss model	Shown, by theoretical analysis
Proof of Space truss model	Shown, by theoretical analysis

Prefabrication Stress Analysis Applications

Proof of Plane Truss Cases

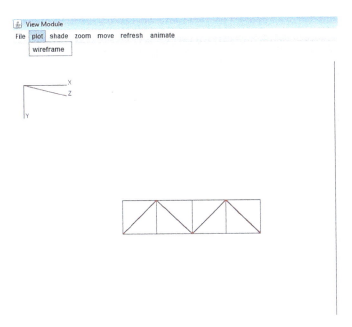

Figure 60. Image of a planar truss.

Figure 60 is a simplified plane truss model, considering half the size of model, with 1 kip load at the center. Note, reactions are equally distributed. Boundary conditions are that one end is hinged the other on a roller.

Simple plane truss data considers one half of the above from stored file 'pltruss.dat' data file. Data structure:

6 (number of nodes)
9 (number of members)
29000.0 elasticity ksi
1 1 2 (member connectivity)
2 2 3
3 3 6
4 4 1

5 4 5

6 5 6

7 1 5

8 2 5

9 3 5

1 10.0 element, member property (area)

2 10.0

3 10.0

4 10.0

5 10.0

6 10.0

7 10.0

8 10.0

9 10.0

1 0.0 0.0 4 (node coordinates)

2 3.0 0.0 4

3 6.0 0.0 4

4 0.0 3.0 4

5 3.0 3.0 4

6 6.0 3.0 4

Boundary condition from boundPtruss.dat file:

2 (number of boundary nodes)

1 1 1 node, (1 indicates restraints) restrain x, restrain y

3 2 1 node, (2 indicates free end, or roller) free x, restrain y

Force data from "force2.dat" file:

1 (number of force conditions)

2 0.0 1.0 (node, fx, fy)

Output Screen 1

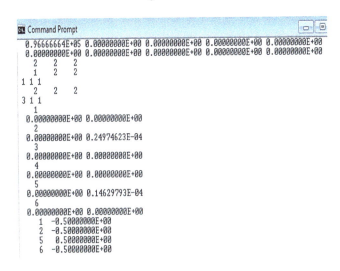

```
C:\ Command Prompt                              [ ] [ ] [ ]
0.96666664E+05 0.00000000E+00 0.00000000E+00 0.00000000E+00 0.00000000E+00
0.00000000E+00 0.00000000E+00 0.00000000E+00 0.00000000E+00 0.00000000E+00
      2      2      2
      1      2      2
1 1 1
      2      2      2
3 2 1
      1
  0.00000000E+00 0.00000000E+00
      2
 -0.51724137E-05 0.30147035E-04
      3
 -0.10344827E-04 0.00000000E+00
      4
 -0.51724141E-05 0.00000000E+00
      5
 -0.51724141E-05 0.19802208E-04
      6
 -0.51724141E-05 0.00000000E+00
      1  -0.30994720E-07
      2  -0.50000000E+00
      5  -0.28609925E-07
      6  -0.50000006E+00
```

Case 1: one end restrained as hinged, other end on roller

```
C:\ Command Prompt                              [ ] [ ]
0.96666664E+05 0.00000000E+00 0.00000000E+00 0.00000000E+00 0.00000000E+00
0.00000000E+00 0.00000000E+00 0.00000000E+00 0.00000000E+00 0.00000000E+00
      2      2      2
      1      2      2
1 1 1
      2      2      2
3 1 1
      1
  0.00000000E+00 0.00000000E+00
      2
  0.00000000E+00 0.24974623E-04
      3
  0.00000000E+00 0.00000000E+00
      4
  0.00000000E+00 0.00000000E+00
      5
  0.00000000E+00 0.14629793E-04
      6
  0.00000000E+00 0.00000000E+00
      1  -0.50000000E+00
      2  -0.50000000E+00
      5   0.50000000E+00
      6  -0.50000000E+00
```

Case 2: same case as 1 except with two ends hinged

Boundary condition from boundPtruss.dat

2

1 1 1 node, restrain x, restrain y

3   1 1

Output Screen 2

Case 3: third case full truss panel

   Coordinates
   1 0.0 0.0 4
   2 3.0 0.0 4
   3 6.0 0.0 4
   4 0.0 3.0 4
   5 3.0 3.0 4
   6 6.0 3.0 4
   7 9.0 0.0 4
   8 9.0 3.0 4
   9 12.0 0.0 4
   10 12.0 3.0 4
   Connectivity and property
   10 (nodes)
   17 (elements)
   29000.0 E (value of elasticity)
   1 1 2 (element connectivity)
   2 2 3
   3 3 6
   4 4 1
   5 4 5
   6 5 6
   7 1 5
   8 2 5
   9 3 5
   10 3 7
   11 6 8
   12 3 8
   13 7 8
   14 7 9
   15 8 10
   16 9 10

17 8 9
    1 10.0
    2 10.0
    3 10.0
    4 10.0
    5 10.0
    6 10.0
    7 10.0
    8 10.0
    9 10.0
    10 10.0
    11 10.0
    12 10.0
    13 10.0
    14 10.0
    15 10.0
    16 10.0
    17 10.0

Force data
1
3 0.0 1.0

Boundary data
2
1 1 1
9 2 1

End of data

```
C:\  Command Prompt

     1
 0.00000000E+00  0.00000000E+00
     2
-0.51724073E-05  0.35319430E-04
     3
-0.10344815E-04  0.60294045E-04
     4
-0.20689637E-04  0.00000000E+00
     5
-0.20689637E-04  0.35319430E-04
     6
-0.10344812E-04  0.60294045E-04
     7
-0.15517220E-04  0.35319423E-04
     8
 0.13669127E-10  0.35319423E-04
     9
-0.20689627E-04  0.00000000E+00
    10
 0.13934122E-10  0.00000000E+00
     1   -0.59084118E-06
     2   -0.50000000E+00
    17   -0.22123459E-07
    18   -0.49999958E+00
```

Case 3 output: one end hinged, another end on roller on full planar truss

Case 4: hinged case

2

1 1 1

9 1 1

```
 Command Prompt
     1
  0.00000000E+00 0.00000000E+00
     2
 -0.30389629E-12 0.24974619E-04
     3
 -0.60779257E-12 0.49949242E-04
     4
 -0.10344826E-04 0.00000000E+00
     5
 -0.10344826E-04 0.24974619E-04
     6
  0.75021369E-12 0.49949238E-04
     7
 -0.47785230E-12 0.24974612E-04
     8
  0.10344827E-04 0.24974612E-04
     9
  0.00000000E+00 0.00000000E+00
    10
  0.10344827E-04 0.00000000E+00
      1   -0.49999997E+00
      2   -0.50000000E+00
     17    0.49999973E+00
     18   -0.49999970E+00
```

Case 4: full planar truss from image above (figure 60); 2 ends hinged, full planar truss

The above image is a screen capture of output for a plane truss model. Examination of the reactions show the program is working correctly for static case; thus, it will be valid for the dynamic case of plane truss as a boom of the crane model.

We develop a space truss example from the plane truss geometry, developed above. Below is an addition to that geometry. Elements are enclosed in circle to differentiate from nodes.

Space Truss

Geometry

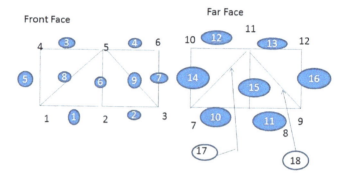

Figure 61. Segment of space truss.

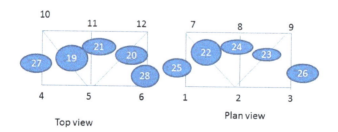

Figure 62. Segment of space truss plan views.

Data:

12

28

0.29000000E+05 psi (or in ksi, 29000)

1 1 2

2 2 3

3 4 5

4 5 6

5 1 4

6 2 5

7 3 6

8 1 5

9 3 5

10 7 8

11 8 9

12 10 11

13 11 12

14 7 10

15 8 11

16 9 12

17 7 11

18 9 11

19 10 5

20 5 12

21 5 11

22 7 2

23 2 9

24 2 8

25 1 7

26 3 9

27 4 10

28 6 12

    1 10.

    2 10.

    3 10.

4 10.
5 10.
6 10.
7 10.
8 10.
9 10.
10 10.
11 10.
12 10.
13 10.
14 10.
15 10.
16 10.
17 10.
18 10.
19 10.
20 10.
21 10.
22 10.
23 10.
24 10.
25 10.
26 10.
27 10.
28 10.
1 0. 0. 0.
2 3. 0. 0.
3 6. 0. 0.
4 0. 3. 0.
5 3. 3. 0.
6 6. 3. 0.
7 0. 0. 4.
8 3. 0. 4.
9 6. 0. 4.
10 0. 3. 4.

11 3. 3. 4.

12 6. 3. 4.

Load data

2

5 0.0   1.0 0.0

11 0.0 1.0 0.0

Boundary conditions: 4

1 1 1 1 (hinged node and three boundary condition for that node)

3 2 1 1 roller (2), fixity in y (1) and z (1) direction

7 1 1 1

9 2 1 1

Results

```
C:\ Command Prompt
0.00000000E+00 0.00000000E+00 0.00000000E+00
   8
0.00000000E+00 0.42426398E+00 0.00000000E+00
   9
0.00000000E+00 0.00000000E+00 0.00000000E+00
  10
0.00000000E+00 0.00000000E+00 0.00000000E+00
  11
0.00000000E+00 0.42426398E+00 0.00000000E+00
  12
0.00000000E+00 0.00000000E+00 0.00000000E+00
   1   -0.49999997E+00
   2   -0.49999997E+00
   3    0.00000000E+00
   7    0.49999997E+00
   8   -0.49999997E+00
   9    0.00000000E+00
  19   -0.49999997E+00
  20   -0.49999997E+00
  21    0.00000000E+00
  25    0.49999997E+00
  26   -0.49999997E+00
  27    0.00000000E+00
```

Screen shot of output: sum of verticals (2k) verified with reactions.

Numbers in the lower half of the image indicate resultants for that degree of freedom (i.e., number 27 is the associated degree of freedom for node 9. 25, 26, 27 are three degrees of freedom for node 9) and indicative of forces Fx, Fy, Fz at that node.

Output Screen 3

This output screen is for the same model but with change in boundary conditions.

```
Command Prompt
 0.00000000E+00 0.00000000E+00 0.00000000E+00
    8
-0.14999995E+00 0.57426399E+00-0.11249994E+00
    9
-0.29999989E+00 0.00000000E+00 0.00000000E+00
   10
-0.14999999E+00 0.00000000E+00 0.35512295E-01
   11
-0.14999998E+00 0.57426399E+00 0.35512336E-01
   12
-0.14999999E+00 0.00000000E+00 0.35512377E-01
    1   -0.24761505E-06
    2   -0.50000006E+00
    3    0.00000000E+00
    7    0.68801121E-07
    8   -0.49999997E+00
    9    0.00000000E+00
   19   -0.18801040E-06
   20   -0.50000000E+00
   21   -0.33915050E-07
   25    0.24761505E-06
   26   -0.50000006E+00
   27   -0.10251995E-07
```

Hinged case

Boundary conditions

4

1 1 1 1

3 1 1 1

7 1 1 1

9 1 1 1

Results shown above.

Plane Truss to Model Geometry

Model arm: mesh and truss element (simplified to plane truss and plane stress elements)

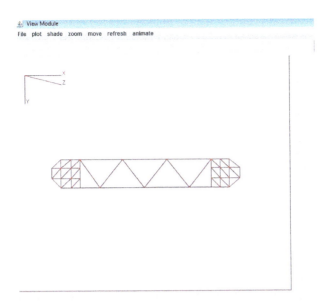

Figure 63. Model with truss elements and plane stress elements (conceptual for model arm).

(Data for the above model is saved in appendix E.) Model is approximated with plane truss and plane stress elements. Shown above.

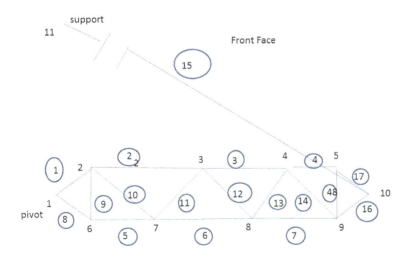

Figure 64. Planar Truss mapping of model geometry.

10 (excluded node 11, member 48 is recorded as 18)
18
29000.0
1 1 2
2 2 3
3 3 4
4 4 5
5 6 7
6 7 8
7 8 9
8 1 6
9 2 6
10 2 7
11 7 3
12 3 8
13 8 4
14 4 9
15 11 10 not a member for this case
16 9 10
17 5 10
18 5 9
    1 10.0
    2 10.0
    3 10.0
    4 10.0
    5 10.0
    6 10.0
    7 10.0
    8 10.0
    9 10.0
    10 10.0
    11 10.0
    12 10.0
    13 10.0
    14 10.0

    15 10.0
    16 10.0
    17 10.0
    18 10.0

Coordinates
1 0. 3. 0.
2 3. 6. 0.
3 6. 6. 0.
4 9. 6. 0.
5 12. 6. 0.
6 3. 0. 0.
7 6. 0. 0.
8 9. 0. 0.
9 12. 0. 0.
10 15. 3. 0.

Force
1
3 0.0 1.0 (load in kips matching elasticity as ksi)

Boundary condition
2
1 1 1
10   2 1

Plane Truss Output Screen
for Model Geometry

```
▣ Command Prompt
 0.00000000E+00  0.00000000E+00
    2
-0.91958582E-05  0.18180144E-04
    3
-0.29159842E-05  0.52799307E-04
    4
 0.12949250E-05  0.31607236E-04
    5
 0.34368688E-05  0.10269212E-04
    6
 0.35477344E-05  0.12119202E-04
    7
 0.51726323E-06  0.40385512E-04
    8
-0.56166568E-05  0.39883103E-04
    9
-0.96816111E-05  0.14553098E-04
   10
-0.77401302E-06  0.00000000E+00
    1    -0.72053701E+00
    2    -0.60000008E+00
   19    -0.30644856E-07
   20    -0.40282169E+00
```

Results show vertical reactions match up with downward load of 1k, with added node 11 for support point of string.

11

18

29000.0

1 1 2

2 2 3

3 3 4

4 4 5

5 6 7

6 7 8

7 8 9

8 1 6

9 2 6

10 2 7

11 7 3

12 3 8
13 8 4
14 4 9
15 11 10
16 9 10
17 5 10
18 5 9
    1 10.0
    2 10.0
    3 10.0
    4 10.0
    5 10.0
    6 10.0
    7 10.0
    8 10.0
    9 10.0
    10 10.0
    11 10.0
    12 10.0
    13 10.0
    14 10.0
    15 10.0
    16 10.0
    17 10.0
    18 10.0

Coordinates
1 0. 3. 0.
2 3. 6. 0.
3 6. 6. 0.
4 9. 6. 0.
5 12. 6. 0.
6 3. 0. 0.
7 6. 0. 0.
8 9. 0. 0.
9 12. 0. 0.

10 15. 3. 0.
11 0.   20.0 0.0

Boundary condition
2
1 (node) 1 1
11(node) 1 1

Output Screen for Case 2

Mapping to model geometry with results

```
CIV  Command Prompt
 -0.16944256E-04  0.30885582E-04
      3
 -0.89118030E-05  0.80839047E-04
      4
 -0.29483199E-05  0.74981275E-04
      5
  0.94619531E-06  0.65472406E-04
      6
  0.24715380E-04  0.28329800E-04
      7
  0.23437491E-04  0.68425259E-04
      8
  0.19056153E-04  0.83257146E-04
      9
  0.16743783E-04  0.73261443E-04
     10
  0.17902954E-04  0.71413815E-04
     11
  0.00000000E+00  0.00000000E+00
      1    -0.35294098E+00
      2    -0.59999979E+00
     21     0.35294074E+00
     22    -0.39999950E+00
```

Noted that vertical reactions at node 1 and 11 sum up to 1k load at node 3.

Case 3: moved load to node 10

```
Command Prompt
-0.19375346E-04  0.32283981E-04
    3
-0.14811458E-04  0.71413800E-04
    4
-0.10247574E-04  0.11054362E-03
    5
-0.56836943E-05  0.14054570E-03
    6
 0.54320455E-04  0.41411786E-04
    7
 0.58884365E-04  0.71413815E-04
    8
 0.63448279E-04  0.11054365E-03
    9
 0.68012181E-04  0.14967348E-03
   10
 0.48636753E-04  0.18195756E-03
   11
 0.00000000E+00  0.00000000E+00
    1    -0.88235360E+00
    2     0.11602036E-05
   21     0.88235217E+00
   22    -0.99999911E+00
```

Noted that vertical reactions at node 1 and 11 sum up to 1k load at node 10

Case with change in boundary condition

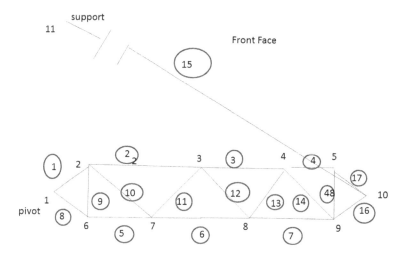

Figure 65. Space truss: view 1.

Proof on Space truss cases

Space truss geometry for model

Model input

    1 1 2
    2 2 3
    3 3 4
    4 4 5
    5 6 7
    6 7 8
    7 8 9
    8 1 6
    9 2 6
    10 2 7
    11 7 3
    12 3 8
    13 8 4
    14 4 9

15 11 10

16 9 10

17 5 10

48 5 9 (**member 48 should be 18 for plane stress**)

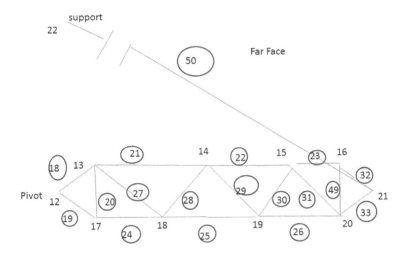

Figure 66. Far face.

18 12 13

19 12 17

20 13 17

21 13 14

22 14 15

23 15 16

24 17 18

25 18 19

26 19 20

27 13 18

28 18 14

29 14 19

30 19 15

31 15 20

32 16 21

33 20 21
49 16 20
50 21 22

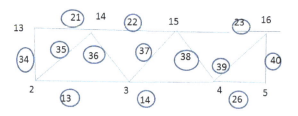

Figure 67. Top face.

34 2 13
35 2 14
36 14 3
37 3 15
38 15 4
39 4 16
40 5 16

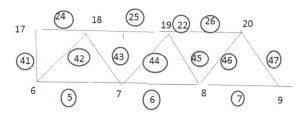

Figure 68. Bottom face.

41 6 17
42 6 18
43 18 7
44 7 19
45 19 8
46 8 20
47 20 9

Proof on Model as Space Truss

From file 'sptruss.txt', combined input:

22
50
0.29000000E+05
1 1 2
2 2 3
3 3 4
4 4 5
5 6 7
6 7 8
7 8 9
8 1 6
9 2 6
10 2 7
11 7 3
12 3 8
13 8 4
14 4 9
15 11 10
16 9 10
17 10 5
18 12 13
19 12 17
20 13 17
21 13 14

22 14 15
23 15 16
24 17 18
25 18 19
26 19 20
27 13 18
28 18 14
29 14 19
30 19 15
31 15 20
32 16 21
33 20 21
34 2 13
35 2 14
36 14 3
37 3 15
38 15 4
39 4 16
40 5 16
41 6 17
42 6 18
43 18 7
44 7 19
45 19 8
46 8 20
47 20 9
48 5 9
49 16 20
50 21 22
   1 10.
   2 10.
   3 10.
   4 10.
   5 10.
   6 10.

7 10.
8 10.
9 10.
10 10.
11 10.
12 10.
13 10.
14 10.
15 10.
16 10.
17 10.
18 10.
19 10.
20 10.
21 10.
22 10.
23 10.
24 10.
25 10.
26 10.
27 10.
28 10.
29 10.
30 10.
31 10.
32 10.
33 10.
34 10.
35 10.
36 10.
37 10.
38 10.
39 10.
40 10.
41 10.

42 10.
43 10.
44 10.
45 10.
46 10.
47 10.
48 10.
49 10.
50 10.
1 0. 3. 0.
2 3. 6. 0.
3 6. 6. 0.
4 9. 6. 0.
5 12. 6. 0.
6 3. 0. 0.
7 6. 0. 0.
8 9. 0. 0.
9 12. 0. 0.
10 15. 3. 0.
11 0. 20. 0.
12 0. 3. 4.
13 3. 6. 4.
14 6. 6. 4.
15 9. 6. 4.
16 12. 6. 4.
17 3. 0. 4.
18 6. 0. 4.
19 9. 0. 4.
20 12. 0. 4.
21 15. 3. 4.
22 0. 20. 4.

Force data force3.dat
2
3 0.0 1.0 0.0
14 0.0 1.0 0.0

Boundary condition
From boundsptruss.dat
6
1 1 1 1
10 1 1 1
11 1 1 1
12 1 1 1
21 1 1 1
22 1 1 1

Output Screen

```
Command Prompt
 -0.26617336E+00 0.42393523E+00 0.64451396E+00
  21
 0.00000000E+00 0.00000000E+00 0.00000000E+00
  22
 0.00000000E+00 0.00000000E+00 0.00000000E+00
   1    -0.28150931E-01
   2    -0.59999955E+00
   3     0.00000000E+00
  28     0.28150771E-01
  29    -0.39999992E+00
  30     0.00000000E+00
  31     0.00000000E+00
  32     0.00000000E+00
  33     0.00000000E+00
  34    -0.28150614E-01
  35    -0.59999985E+00
  36     0.00000000E+00
  61     0.28151175E-01
  62    -0.39999953E+00
  63     0.00000000E+00
  64     0.00000000E+00
  65     0.00000000E+00
  66     0.00000000E+00
```

Analysis shows vertical reactions match up downward loads. Displacements at nodes listed below:
1
0.00000000E+00 0.00000000E+00 0.00000000E+00
2
-0.25882655E+00 0.52532822E+00 0.12963085E+00
3
-0.74604005E-01 0.15305470E+01 -0.85360520E-02
4

0.49618538E-01 0.91724098E+00 -0.10170297E+00

5

0.11384117E+00 0.29549009E+00 -0.14986986E+00

6

0.11115880E+00 0.35377359E+00 0.36151403E+00

7

0.25381478E-01 0.11705472E+01 0.42584738E+00

8

-0.15039581E+00 0.11572410E+01 0.55768079E+00

9

-0.26617312E+00 0.42393532E+00 0.64451396E+00

10

0.00000000E+00 0.00000000E+00 0.00000000E+00

11

0.00000000E+00 0.00000000E+00 0.00000000E+00

12

0.00000000E+00 0.00000000E+00 0.00000000E+00

13

-0.25882658E+00 0.52532822E+00 0.12963088E+00

14

-0.74604079E-01 0.15305470E+01 -0.85360324E-02

15

0.49618464E-01 0.91724092E+00 -0.10170294E+00

16

0.11384108E+00 0.29549000E+00 -0.14986986E+00

17

0.11115850E+00 0.35377353E+00 0.36151403E+00

18

0.25381105E-01 0.11705470E+01 0.42584738E+00

19

-0.15039618E+00 0.11572406E+01 0.55768073E+00

20

-0.26617336E+00 0.42393523E+00 0.64451396E+00

21

0.00000000E+00 0.00000000E+00 0.00000000E+00

22
0.00000000E+00 0.00000000E+00 0.00000000E+00

Case 2 has the same loads but changes in boundary condition. Summation in vertical reaction balance downward loads:

6
1 1 1 1
10 2 2 1
11 1 1 1
12 1 1 1
21 2 2 1
22   1 1 1

This is another condition for a run-time check with case 2 data:

Force conditions
From force3.dat
2
10 0.0 1.0 0.0
21 0.0 1.0 0.0

Boundary condition boundsptruss.dat
6
1 1 1 1
11 1 1 1
10 2 2 1
21 2 2 1
12 1 1 1
22 1 1 1

Results and output screen

```
Command Prompt
 0.48557135E+00  0.21245859E+01-0.18033405E+00
  21
 0.51918697E+00  0.20710051E+01  0.00000000E+00
  22
 0.00000000E+00  0.00000000E+00  0.00000000E+00
   1    -0.35294154E+00
   2    -0.59999925E+00
   3     0.00000000E+00
  28     0.98575725E-07
  29     0.94678296E-06
  30     0.00000000E+00
  31     0.35294127E+00
  32    -0.40000010E+00
  33     0.00000000E+00
  34    -0.35294130E+00
  35    -0.59999949E+00
  36     0.00000000E+00
  61    -0.81367091E-07
  62     0.57010436E-06
  63     0.00000000E+00
  64     0.35294145E+00
  65    -0.40000030E+00
  66     0.00000000E+00
```

Displacements output (Ux, Uy, Uz are displacements in x, y, and z directions)

	Ux	Uy	Uz
1	0.00000000E+00	0.00000000E+00	0.00000000E+00
2	0.29354811E-04	0.19892563E-03	-0.12427000E-01
3	-0.80360252E-04	0.98149991E-03	-0.11601935E-01
4	-0.16565922E-03	0.15964867E-02	-0.10805533E-01
5	-0.27411632E-03	0.20937016E-02	-0.99900467E-02
6	-0.43834207E-03	0.24476962E-03	-0.23462495E-03
7	-0.45744728E-03	0.10006280E-02	-0.89750596E-04
8	-0.45349644E-03	0.16128063E-02	0.35479294E-04

9
-0.47393981E-03 0.21069292E-02 0.19065422E-03
10
-0.46482935E-03 0.22346748E-02 0.00000000E+00
11
0.00000000E+00 0.00000000E+00 0.00000000E+00
12
0.00000000E+00 0.00000000E+00 0.00000000E+00
13
0.29286703E-04 0.19893928E-03 -0.12427001E-01
14
-0.80603750E-04 0.98179292E-03 -0.11601936E-01
15
-0.16595557E-03 0.15967836E-02 -0.10805534E-01
16
-0.27439487E-03 0.20939345E-02 -0.99900467E-02
17
-0.43851885E-03 0.24482736E-03 -0.23462495E-03
18
-0.45765887E-03 0.10009141E-02 -0.89750589E-04
19
-0.45372863E-03 0.16130973E-02 0.35479294E-04
20
-0.47418659E-03 0.21071599E-02 0.19065422E-03
21
-0.46508538E-03 0.22349320E-02 0.00000000E+00
22
0.00000000E+00 0.00000000E+00 0.00000000E+00

Member-end forces
1
    1-0.27742374E+01 0.00000000E+00 0.00000000E+00
    2 0.27742374E+01 0.00000000E+00 0.00000000E+00
    2
    2 0.10605789E+01 0.00000000E+00 0.00000000E+00
    3-0.10605789E+01 0.00000000E+00 0.00000000E+00

3
3 0.62569916E+00 0.00000000E+00 0.00000000E+00
4-0.62569916E+00 0.00000000E+00 0.00000000E+00
4
4 0.10484186E+01 0.00000000E+00 0.00000000E+00
5-0.10484186E+01 0.00000000E+00 0.00000000E+00
5
6 0.18468349E+00 0.00000000E+00 0.00000000E+00
7-0.18468349E+00 0.00000000E+00 0.00000000E+00
6
7-0.23518738E+00 0.00000000E+00 0.00000000E+00
8 0.23518738E+00 0.00000000E+00 0.00000000E+00
7
8 0.19761927E+00 0.00000000E+00 0.00000000E+00
9-0.19761927E+00 0.00000000E+00 0.00000000E+00
8
1 0.67610202E+01 0.00000000E+00 0.00000000E+00
6-0.67610202E+01 0.00000000E+00 0.00000000E+00
9
2 0.10724655E+01 0.00000000E+00 0.00000000E+00
6-0.10724655E+01 0.00000000E+00 0.00000000E+00
10
2 0.66518393E+01 0.00000000E+00 0.00000000E+00
7-0.66518393E+01 0.00000000E+00 0.00000000E+00
11
7 0.10937452E-01 0.00000000E+00 0.00000000E+00
3-0.10937452E-01 0.00000000E+00 0.00000000E+00
12
3 0.51750302E+01 0.00000000E+00 0.00000000E+00
8-0.51750302E+01 0.00000000E+00 0.00000000E+00
13
8-0.17812420E-01 0.00000000E+00 0.00000000E+00
4 0.17812420E-01 0.00000000E+00 0.00000000E+00
14
4 0.42238889E+01 0.00000000E+00 0.00000000E+00

9-0.42238889E+01 0.00000000E+00 0.00000000E+00
15
11 0.26556857E+01 0.00000000E+00 0.00000000E+00
10-0.26556857E+01 0.00000000E+00 0.00000000E+00
16
9-0.27317575E+00 0.00000000E+00 0.00000000E+00
10 0.27317575E+00 0.00000000E+00 0.00000000E+00
17
10 0.10513941E+01 0.00000000E+00 0.00000000E+00
5-0.10513941E+01 0.00000000E+00 0.00000000E+00
18
12-0.27723942E+01 0.00000000E+00 0.00000000E+00
13 0.27723942E+01 0.00000000E+00 0.00000000E+00
19
12 0.67632127E+01 0.00000000E+00 0.00000000E+00
17-0.67632127E+01 0.00000000E+00 0.00000000E+00
20
13 0.10732234E+01 0.00000000E+00 0.00000000E+00
17-0.10732234E+01 0.00000000E+00 0.00000000E+00
21
13 0.10622745E+01 0.00000000E+00 0.00000000E+00
14-0.10622745E+01 0.00000000E+00 0.00000000E+00
22
14 0.62620807E+00 0.00000000E+00 0.00000000E+00
15-0.62620807E+00 0.00000000E+00 0.00000000E+00
23
15 0.10482466E+01 0.00000000E+00 0.00000000E+00
16-0.10482466E+01 0.00000000E+00 0.00000000E+00
24
17 0.18502030E+00 0.00000000E+00 0.00000000E+00
18-0.18502030E+00 0.00000000E+00 0.00000000E+00
25
18-0.23499009E+00 0.00000000E+00 0.00000000E+00
19 0.23499009E+00 0.00000000E+00 0.00000000E+00
26

19 0.19776025E+00 0.00000000E+00 0.00000000E+00
20-0.19776025E+00 0.00000000E+00 0.00000000E+00
27
13 0.66539502E+01 0.00000000E+00 0.00000000E+00
18-0.66539502E+01 0.00000000E+00 0.00000000E+00
28
18 0.11016568E-01 0.00000000E+00 0.00000000E+00
14-0.11016568E-01 0.00000000E+00 0.00000000E+00
29
14 0.51749430E+01 0.00000000E+00 0.00000000E+00
19-0.51749430E+01 0.00000000E+00 0.00000000E+00
30
19-0.17772846E-01 0.00000000E+00 0.00000000E+00
15 0.17772846E-01 0.00000000E+00 0.00000000E+00
31
15 0.42232761E+01 0.00000000E+00 0.00000000E+00
20-0.42232761E+01 0.00000000E+00 0.00000000E+00
32
16 0.11140465E+01 0.00000000E+00 0.00000000E+00
21-0.11140465E+01 0.00000000E+00 0.00000000E+00
33
20-0.27317461E+00 0.00000000E+00 0.00000000E+00
21 0.27317461E+00 0.00000000E+00 0.00000000E+00
34
2-0.71834773E-04 0.00000000E+00 0.00000000E+00
13 0.71834773E-04 0.00000000E+00 0.00000000E+00
35
2 0.20892072E+01 0.00000000E+00 0.00000000E+00
14-0.20892072E+01 0.00000000E+00 0.00000000E+00
36
14-0.83152205E-04 0.00000000E+00 0.00000000E+00
3 0.83152205E-04 0.00000000E+00 0.00000000E+00
37
3 0.16271993E+01 0.00000000E+00 0.00000000E+00
15-0.16271993E+01 0.00000000E+00 0.00000000E+00

38
15-0.83219260E-04 0.00000000E+00 0.00000000E+00
4 0.83219260E-04 0.00000000E+00 0.00000000E+00
39
4 0.20654645E+01 0.00000000E+00 0.00000000E+00
16-0.20654645E+01 0.00000000E+00 0.00000000E+00
40
5-0.67315996E-05 0.00000000E+00 0.00000000E+00
16 0.67315996E-05 0.00000000E+00 0.00000000E+00
41
6 0.12619421E-06 0.00000000E+00 0.00000000E+00
17-0.12619421E-06 0.00000000E+00 0.00000000E+00
42
6 0.36693281E+00 0.00000000E+00 0.00000000E+00
18-0.36693281E+00 0.00000000E+00 0.00000000E+00
43
18 0.73981937E-06 0.00000000E+00 0.00000000E+00
7-0.73981937E-06 0.00000000E+00 0.00000000E+00
44
7-0.68772763E-01 0.00000000E+00 0.00000000E+00
19 0.68772763E-01 0.00000000E+00 0.00000000E+00
45
19-0.98050805E-07 0.00000000E+00 0.00000000E+00
8 0.98050805E-07 0.00000000E+00 0.00000000E+00
46
8 0.39302117E+00 0.00000000E+00 0.00000000E+00
20-0.39302117E+00 0.00000000E+00 0.00000000E+00
47
20-0.90454705E-07 0.00000000E+00 0.00000000E+00
9 0.90454705E-07 0.00000000E+00 0.00000000E+00
48
5 0.36141691E+00 0.00000000E+00 0.00000000E+00
9-0.36141691E+00 0.00000000E+00 0.00000000E+00
49
16 0.36136034E+00 0.00000000E+00 0.00000000E+00

20-0.36136034E+00 0.00000000E+00 0.00000000E+00

50

21-0.15668010E+01 0.00000000E+00 0.00000000E+00

22 0.15668010E+01 0.00000000E+00 0.00000000E+00

These member-end reactions are expected to match up with calibrated strain gauge when affixed to any member and when activated with wireless device from experimental stress analysis.

Simulation: Plane Truss and Space Truss

To demonstrate integration, we isolate a component of truss assembly and perform a finite element analysis with a unit load downward at two ends of the truss. This analysis selects typically 10 instances of boom position as it moves through 10 angles of lift for a unit load applied at the end; thus, this simulates the truss outline of the fabricated physical model. Other component of the load goes into the string. The black line is an outline displaying the movement. It is symbolic of the movement of crane arm.

Later in the section it is related to experimental stress analysis on a postfabricated model; however, to retain the crane model as a nondestructive sample, we apply this concept on a beam model made with aluminum for a static case and confirm the relation between theoretic analysis and experimental analysis.(see part 7: Wireless Communication).

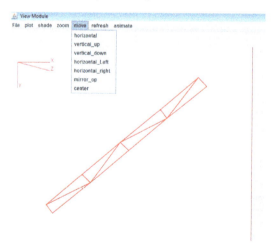

Figure 69. Image of truss component of the model (two end plates not shown).

Coupling

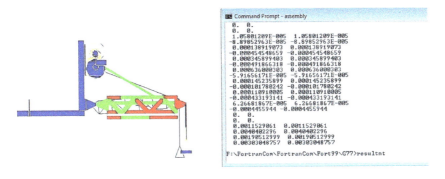

Figure 70. Run-time simulation with right side showing analysis.

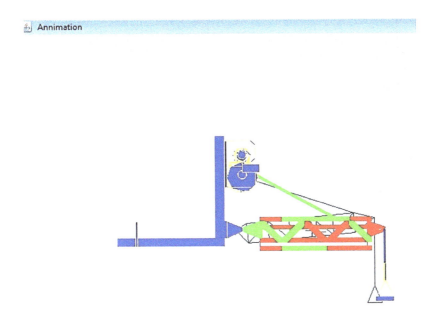

Figure 71. Stress analysis members in red crossing threshold limits.

In figure 71, the image shows some members passing (green) and some members failing (red). Items in blue are not analyzed.

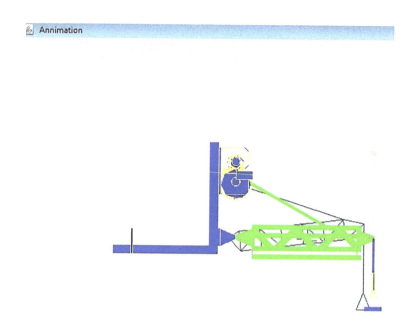

Figure 72. Graphical display of results.

Figure 72 shows all members passing and shaded in green color. Items in blue are not analyzed. The screen image displays a run-time simulation between animation and stress analysis on selected members of truss elements and color coded for passing.

Constraints with Lagrange Multipliers external

Lagrange multipliers are additional parameters to satisfy compatibility for two-dimensional problems such as on plate bending problems. Lagrange multipliers are constraining numbers that are applied to a system of equation. These form topics in more advanced finite element analysis. Reader may want to refer to the cited reference in footnote..[29]

---

29. For more information, see Tong and Rossettos, *Introduction to Finite Elements ,231*

Method of Solution for FEA

We adopt a method of Gaussian elimination on half the bandwidth .[30] An example is included on beam and plate element of plane stress/strain and plate bending (figure 59), as well as on a space truss element via animation, coupled with FEA. The program in the list of batch file uses this method of solution. Alternatively writers program on fully populated matrix gives the same results.

Load vector is developed when load is provided. It is superimposed with displacement specified.

Limitations both cannot be specified at the same node.

The solution to a problem depends on the proper **boundary values**; thus, there is a need to specify the proper anchorage and boundary conditions.

Database for Member Selection

The database for member selection is described in part 2.

Conclusion

In this chapter, we added a concept of run-time simulation with the FE method on a component of the model and linking with animation. The stiffness matrix changes for each position; thus, simulation produces new results as the model arm takes up a different position. This simulation is further related with postfabrication stress analysis. The concept is projected to the crane model. We highlighted the basics of FEA and sent pointers to advanced FEA methods. We presented screen output as proof for various conditions. This aspect with respect to fabricated model is the distinctiveness of this manual.

---

30. Jenkins, *Matrix and Digital Computer Methods in Structural Analysis.*

PART 5:
Iterative Methods

The Gram-Schmidt Orthogonalization[31]

Objective

Our objective is to explore iterative methods and apply them to the model.

Background

This is a precursor section for dynamic analysis where orthogonal vectors are used in finding the response. The algorithm is implemented from the reference , in the computer program (see footnote 33). This process will build independent vectors that are orthogonal to each other. We use the property of dot product where two vectors are orthogonal if their dot product is zero. In other words: a.b = ab cos (theta) is zero when theta is 90 degrees.

This algorithm is cyclic in nature and involving dot product and normalization of vectors. Orthogonalization is with respect to mass matrix; thus, if V3 is a vector, then a corresponding dot product of vectors, which is orthogonal between two vectors V2 and V3 with respect mass matrix, is given by this logic below.

31. Albert Boggess, and Francis J. Narcowich, *First Course in Wavelets with Fourier Analysis* (London, England: Pearson, 2001), 19–20.(**with permission**) Mass is not cited in this reference, but I've added this based on other references, namely by Biggs and Meirovitch.

Logic: <e2,e1> e1 ,where e1 = V3/||V3|| , e1 is a unit vector
    E2= {V2 trans[M][V3/||V3||]}.e1
    E2 Orthogonal projection of V2 on to V3 vector
    V2 = E2 + P
    P= V2-E2 where P is new orthogonal vector
    Next cycle
    e2= e2/|e2| = P/||P||
We repeat the process with another V4 vector and make it orthogonal to V2 and V3.

After finding the most effective vectors through participation factors, they are further multiplied to the force matrix and pre- and post-multiplied to damping matrix, stiffness matrix, and mass matrix. A new eigenvalue problem is solved on this reduced matrix.

We start from an assumed vector and another vector. In this case, we take the displacement produced by the static portion of the dynamic load on the beam model. The next vector generated is a displacement produced by the lumped mass at the joints. The displacement produced by each lumped mass is calculated and made orthogonal to the vectors already in orthogonal state. Based on the heavier participation of a vector, those vectors are chosen. These selected strong vectors are then pre-multiplied and post-multiplied to stiffness and mass matrix. Eigenvalues are then computed on this reduced stiffness and mass matrix. An algorithm to develop orthogonal vectors is shown below:

Pseudocode:
Orthogonal projection of an arbitrary vector V2 in the direction of V3 is given by <dot product V2[M].V3/||V3||>

Next we multiply with the unit vector found in the direction of V3, leading to:

<dot product V2[M]>* V3/||V3|| (V3 known vector)
E1= Unit vector = V3/||V3|| (known vector)
E1 gives a unit vector in the direction of V3
V1 is orthogonal projection of V2 on V3 = [V2(trans)[M]
V3/ ||(V3)|||]*E1

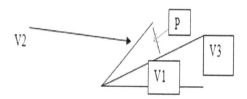

Figure 73. The Gram-Schmidt orthogonal vector diagram.

From the figure 73

$$V1 + P = V2$$

Or P=- V2 – V1

Equation 14

V1 is the orthogonal projection of V2 onto V3 vector.
P is a new orthogonal vector.

This process is cyclic by taking the next displacement produced from the solution of stiffness matrix using another mass (M2). Based on this a new displacement, another orthogonal vector is developed to the new displacement vectors and making it orthogonal to all previous orthogonal vectors (see footnotes 32 and 33). An example of this is demonstrated in the computer program for beam analysis:

Computer program is developed in Fortran
Beam2.for is Fortran program compiled to beam2.exe using g77 compiler.
(G77 beam2.for –o beam2.exe)
We show generating orthogonal vectors from a disturbance on a beam.
We construct orthogonality with respect to mass matrix.
[V](n) transpose [M][Vn-1...1] =0,

This is a practical application. These orthogonal vectors are known as **RITZ_VECTORS.**[32] This algorithm is demonstrated with the following

---

32. Wilson, E. L. and Tetsuji Itoh. "An Eigensolution Strategy for Large Systems." *Computers and Structures* 16, no. 1–4 (1983): 259–265.

output. The model on figure 78 is used and solved with computer program "bsolve2" part of "beam2.exe". Initial vector of displacement is with an assumed 4 lb. load at node 7, subsequently a 1 lb. mass is applied independently on the same model at five points,(nodes 6,5,4,3) in vertical direction sequentially. The **displacement is orthogonalized with each of previous displacement vector starting from first displacement vector.** Orthogonality is checked at the end of computation by applying the dot product on vectors. The output shown below is for 5 vectors and checked for orthogonality of each vector with the first vector and noted to be 0.(V1.Vn..)=0) meeting orthogonality condition. First Vector is the initial displacement.

V1 (V is a symbol for vector)
0.00000000E+00 0.00000000E+00 0.91159260E-02 0.15729442E-01 0.34318782E-01
0.28598987E-01 0.72391182E-01 0.38608626E-01
0.12011573E+00 0.45758385E-01
0.17427507E+00 0.50048221E-01 0.23165178E+00 0.51478177E-01
V2
0.00000000E+00 0.00000000E+00 0.13921801E-03 0.21919061E-03 0.42948645E-03
0.26853426E-03 0.67972747E-03 0.14803234E-03 0.69886533E-03-0.14231545E-03
0.29582341E-03-0.60250895E-03-0.58641564E-03-0.87506667E-03
V3
0.00000000E+00 0.00000000E+00 0.40611059E-04 0.60768478E-04 0.11101185E-03
0.52959196E-04 0.13405284E-03-0.23428096E-04 0.32584401E-04-0.16839267E-03
-0.13648925E-03-0.24454128E-04 0.11171272E-04 0.20911060E-03
V4
0.00000000E+00 0.00000000E+00 0.38863669E-04 0.53403492E-04 0.84863983E-04
0.12687759E-04 0.32115808E-04-0.12214827E-03-0.91211616E-04 0.63758885E-05
0.26865979E-04 0.82650600E-04 0.12451505E-04-0.60546561E-04

V5

0.00000000E+00 0.00000000E+00 0.43812779E-04 0.49463219E-
04 0.47330952E-04

-0.71635237E-04-0.47271100E-04-0.58131677E-05 0.26030579E-
04 0.38084239E-04

-0.48448219E-05-0.33388504E-04-0.21768560E-05 0.20252282E-04

A check on orthogonality is listed below ignoring the round off digits
(such as 10^{-10})
Check Orthogonality with 1 and 2 (V1.V2) = 0.00
Check Orthogonality with 1 and 3 (V1.V3 = 0.00
Check Orthogonality with 1 and 4(V1.V4) = 0.00
Check Orthogonality with 1 and 5 (V1.V5) = 0.00

Conclusion

This method of Gram-Schmidt allows reducing a large matrix to a smaller
matrix by selecting vectors that have a significant contribution to the
problem. Eigenvalues and eigenvectors are then computed from this
smaller problem; however static amplitude of force must be available as the
method is based on this initial value.

PART 6:
Finite Elements and Dynamic Systems

Objective

The objective is to extend finite element and iterative methods to problems of dynamics.

Background

Dynamic systems can be solved using the finite element method by applying Newton's second law of force and accelerations. We demonstrate the application on the model with a simple beam structure.

Dynamic Systems[33]

An example of the dynamic system is shown below.

---

33. For more information, please see Biggs, *Introduction to Structural Dynamics* and Meirovitch, *Elements of Vibration Analysis*.

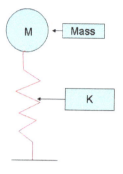

Figure 74. Typical one-degree dynamic system.

Equation for Dynamic System

mX.. + CX. + Kx = F(t)

Where m = mass of the system

C = damping coefficient

K= stiffness

x.. = acceleration

x. = velocity

x= displacement

Figure 75. Dynamic system.

Finite element programs are in Fortran language using the g77 Fortran compiler. It is taken that they are platform independent. Applications to bodies are described with a beam.

Image of Interactive Screen

Figure 76. Typical FEA arrangement.

Eigenvalues and Eigenvectors

A typical equation for eigenvalue problems:

Ax = (lambda) λ x
Where: A is an operator or matrix
      X is the eigenvector
      Lambda is eigenvalue (frequency) associated with
      the eigenvector

An application with the program is demonstrated on a problem cited in Meirovitch.[34] This proves the program output matches up with the same result. It can be concluded that its verification to other structural model will have similar result. This algorithm is applied to the bar model shown in figure 77 where stiffness matrix is produced by finite element method. This is an iterative method, and the method of deflation follows from Meirovitch. For example, an output from Meirovitch[35] is shown but with the application from our program developed in Fortran. A match to the values in the reference text is found; thus, this a proof that matrix deflation method works correctly from our program.

---

34. Leonard Meirovitch, *Elements of Vibration Analysis* (New York: McGraw-Hill, 1975).

35. Meirovitch, *Elements of Vibration Analysis*, 165–166.

Output saved in file named "Out.vec"

Frequency 1: 0.37308733E+00

Vectors: 0.46259845E+00 0.86080585E+00 0.10000000E+01

Frequency 2: 0.13212704E+01

Vectors: -0.10000000E+01 -0.25424451E+00 0.34072676E+00

Example from an output of the beam model showing iterations, mode shape, and frequency.

Now it will be applied to the bar model using the method of deflation.

Figure 77. Cantilever bar.

FEA on a beam model

Figure 78. FEA model for cantilever bar.

Node 1 is modeled with fixed boundary condition.

Lumped mass at six nodes and is shown below. This program uses the stiffness matrix developed from FEA on experimental stress analysis page.

```
F:\FortranCom\FortranCom\Fort99\G77>dyan12_1
 12
0.51096659E-02
0.00000000E+00
0.51096659E-02
0.00000000E+00
0.51096659E-02
0.00000000E+00
0.51096659E-02
0.00000000E+00
0.51096659E-02
0.00000000E+00
0.17032219E-02
0.00000000E+00
```

Mode shapes (eigenvectors) and frequencies (eigenvalues) for the model above are captured from an output.

First value

(Lambda) frequency 0.50378072E+02

Vectors

0.45478205E-01 0.77439673E-01 0.16664469E+00 0.13462727E+00

0.34116381E+00 0.17254617E+00 0.54853050E+00 0.19355038E+00

0.77185582E+00 0.20176456E+00 0.10000000E+01 0.20330997E+00

Normalized vectors

0.54233099E+00 0.92347388E+00 0.19872504E+01 0.16054402E+01
0.40684039E+010.20576260E+01 0.65412672E+01 0.23081027E+01
0.92044383E+01 0.24060575E+01 0.11925075E+02 0.24244865E+01

Second value
(Lambda) frequency 0.30569154E+03

Vectors
0.20885268E+00 0.30357235E+00 0.54403264E+00 0.23646528E+00
0.65008662E+00 -0.72810160E-01 0.36583244E+00 -0.42042733E+00
-0.24522536E+00-0.63297778E+00
-0.10000000E+01 -0.68987547E+00

Normalized vectors
0.25729548E+01 0.37398511E+01 0.67021951E+01 0.29131275E+01
0.80087241E+01-0.89698275E+00 0.45068627E+01-0.51794428E+01-
0.30210470E+01-0.77979523E+01 -0.12319472E+02-0.84989017E+01

Method to get response

With eigenvalues on a decoupled equation, we need to connect the frequency and vectors, and do a time integration to get the time response of the system subjected to external disturbance. Eigenvalues and -vectors only help in finding this response. We limited our scope to finding eigenvalues and eigenvectors. Interested readers may want to follow up in Biggs's book *Introduction to Structural Dynamics or reference in footnote 34.* The response to the system uses mode shape and frequencies with time dependence.

Nonlinear Analysis

Nonlinear analysis is an extension of linear analysis. To this end, an iterative method can be applied and extend the linear analysis. Of the methods available, we have the Newton-Raphson method. This is a separate feature and beyond the scope of the present context.

Conclusion

Our method is a conventional method of computing eigenvalues and eigenvectors. Program results tally with already published problem. The method used here is a method of deflation. Further to this is a time history, which is left as future enhancement. This technique is also applicable to the model, if dynamic simulations are desired, as the stiffness matrix changes for each position of the model arm.

PART 7:
Postfabrication Stress Analysis

The distinctive feature of this segment is the model itself and experiments from the model. Theoretical and experimental results are compared using a beam model as an alternate to the crane model. This is demonstrated with strain gauge capturing the data from a wireless setup. We also explore photoelasticity and LabVIEW options.

Background

Experimental Stress Analysis Framework

There are two methods, namely photoelasticity and strain measurement with strain gauges.

Compatibility in a continuous domain can be viewed by comparison of theoretical results and experimental stress analysis and thus make an error approximation. The process complements numerical analysis of FE.

The model is the basis and projected to prototypes such as bridges or cranes as a real-world application. Two methods for converting analog values to digital values are via LabVIEW and wireless acquisition using MicroStrain devices. We try strain variation with potentiometer, as well as affixing strain gauges to the sample. Samples include aluminum bar, wooden bar, and a plastic bar.

Photoelasticity[36]

Photoelasticity is addressed to encompass the field of experimental stress analysis. This is one of the earlier methods in experimental stress analysis, dating back to early 1900s.

A simplified demo shows the effect of photoelasticity. This is an alternative method of analyzing stresses in a loaded portion of any birefringent material and complements finite elements. The model is viewed under polarized light to illuminate the patterns of stresses. Photoelasticity can only be applied to a model, whereas strain gauge sensors have no such limitations, provided environmental condition permit this. This is a simple presentation; more elaborate exhibits can be seen in Max M. Frocht's, *Photo elasticity.*

Our model is a candidate for photo elasticity. Fringe order and fringe value are the stress parameters that the method gives. We take a small external model as an example and apply compressive load to the sample. Stress field can be seen when the sample is viewed under polarized light.

Sample

Material Plastic

---

36. For more information, see Max M. Frocht, *Photoelasticity* (New York: John Wiley & Sons, 1941).

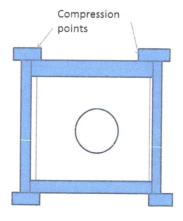

Figure 79. Sample of plastic material.

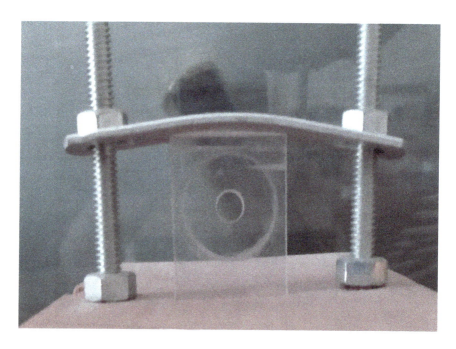

Figure 80. Sample model subjected to compression.

Figure 81. Sample under normal light; no fringes are observed.

Figure 82. Sample under polarized light; highlighted fringes visible are indicative of stress field.

This was and still is one of the valuable methods of visual stress analysis. Mechanical parts are modeled on to a birefringent material, such as plastic, and viewed under polarized light. Contours from numerical finite element analysis under plane stress and visual stress field as above can be compared for conformance with this method.

Conclusion

Photo elasticity is one of the experimental stress analysis methods. We present a concept and direct details to literature on the subject in the cited reference. The model itself is made in plastic and is a candidate for this method. Areas of stress concentration can be visualized by this method.

Alternate method of measuring strains

One alternative method to measuring strains is to use strain gauges. We applied this to a Wheatstone bridge configuration by measuring the change in resistance. The purpose of this method aids in verifying the theoretical results with experiments. We use a general-purpose device **potentiometer** to simulate variation in voltages due to applied loads and label it as a strain sensor.

Experimental Stress Analysis- Noted below is an excerpt from manual on Strain gauge from National Instruments Tutorial Application 078:

"In practice, strain measurements rarely involve quantities larger than a few milli strain (e x 10^{-3}). Therefore, to measure the strain requires accurate measurement of very small changes in resistance. For example, suppose a test specimen undergoes a strain of 500 me. A strain gage with a gage factor of 2 will exhibit a change in electrical resistance of only 2 (500 x 10^{-6}) = 0.1%. For a 120 Ω gage, this is a change of only 0.12 Ω. To measure such small changes in resistance, strain gages are almost always used in a bridge configuration with a voltage excitation source. The general Wheatstone bridge, illustrated in Figure below, consists of four resistive arms with an excitation voltage, V_{EX}, that is applied across the bridge.[37]"

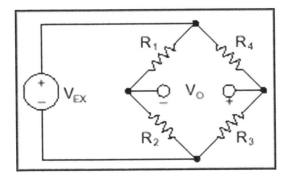

Figure 83. Wheatstone bridge.

The output voltage of the bridge, V_O, is equal to:

$$V_0 = \left[\frac{R_3}{R_3 + R_4} - \frac{R_2}{R_1 + R_2} \right] \bullet V_{EX}$$

From this equation, it is seen that when $R_1/R_2 = R_4/R_3$, the voltage output V_O is zero. Under these conditions, the bridge is said to be balanced. Any change in resistance in any arm of the bridge results in a nonzero output voltage. Therefore, if we replace R_4 in the figure above with an active strain gage, any changes in the strain gage resistance will unbalance the bridge and produce a nonzero output voltage. If the nominal resistance of the strain gage is designated as R_G, then the strain induced by change in resistance, (delta) DR, can be expressed as DR = R_G· (GF) ·e, as defined by the gage factor equation. This assumes that R_1 = $R_{2= R3}$ and R_4 = R_G, the bridge equation above can be rewritten to express V_O/V_{EX} as a function of strain (see "Figure 84. Quarter-Bridge Circuit").

The presence of 1/ (1+GF·e/2) term that indicates the nonlinearity of the quarter-bridge output with respect to strain."

---

37. National Instrument Strain gauge Tutorial Application 078, page 3.

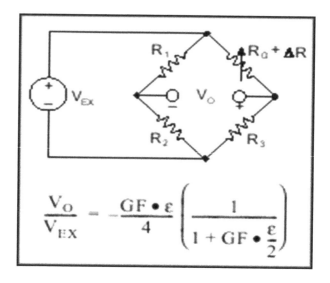

Figure 84. Quarter-Bridge Circuit

Example:

Vex= 5 volts

Vo = 20 mil volts

Then strain e = Vo/Vex * 0.5, where 0.5 = (GF/4) (neglecting nonlinear term)

Stress = E*strain (as per Hooke's law)

Limitations from National Instrument Tutorial Application 078 [38]

> "Ideally, we would like the resistance of the strain gage to change only in response to applied strain. However, strain gage material, as well as the specimen material to which the gage is applied, also responds to changes in temperature. Strain gage manufacturers attempt to minimize sensitivity to temperature by processing the gage material to compensate for the thermal expansion of the specimen material for which the gage is intended. While compensated gages reduce the thermal sensitivity, they do not totally remove it."

---

38. National Instrument Strain gauge Tutorial Application 078, page 2

The models below demonstrate the above statements with visual aids.

In this study, we demonstrate only the variation with experiment and do not dwell on the exact values. Some of the compatibility issues of finite element can be verified and can allow individual to make a prudent judgement on the theoretical values. This method complements the finite element method and can be used as verifiable method of theoretical analysis. To this end we use two approaches:

1. With LabVIEW called a wired system
2. With wireless MicroStrain device provided by LORD MicroStrain

Using self-made model shown in figure 7, we experiment with a quarter-bridge strain gage assembly for collection of data on a beam model. We use breadboard for making connections of quarter-bridge strain gauge.

Setup with LabVIEW

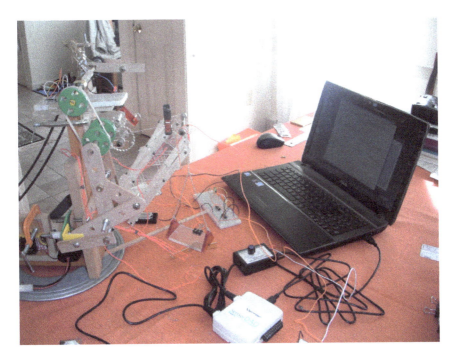

Figure 85. Typical connection to the computer from data acquisition sensor.

A demonstration with strain gauges is shown in figure 96. We make use of LabVIEW from National Instruments as a method to get the strains. Since strain is transformed to voltages, we use an analog sensor to connect to the computer to see its digital response; thus, an analog to digital converter tool is used. A breadboard with LEDs is used as a visual aid for demonstration.

Test the breadboard for bridge circuit

In the figure 86, three LEDs represent the power distribution at input, two red LEDs at resistors, (one at variable resistor R4, one at R2 location), and a green LED at the amplifier node. Potentiometer being the variable resistor is shown off the breadboard to the left.

Figure 86. Switch is open no power in the system.

At switch closed, equal intensity on two LEDs is indicative of a balanced condition (figure 87). Upper green LED at the amplifier is unlit (as no current flows through it).

Case 1 uses Potentiometer without connection to LabVIEW, for testing circuit only. This is demonstrated to the limit of breadboard circuitry by noticing the intensity of LED.

For a practical display, we add an ammeter (substituted with green LED) to view the change in resistance by viewing the intensity of LED lights (figure 88). Two instances are noticed. When there is a balanced condition, no current is indicated in the ammeter (green LED is off) and two red LEDs have the same intensity of light. The resistance of the potentiometer is 120 ohms according to the balance condition equation because other resistances are 120 ohms.

Figure 87. Display with two lighted LEDS showing balanced condition.

However, when we change the resistance on the **potentiometer**, the intensity of the two red LED is different and the ammeter (substituted with green LED indicator) shows a current flow. This is due to the unbalance created by changing the resistance of the potentiometer and is indicative of strain created in the system that would be used typically to compute stress.

Figure 88. Green LED indicative of unbalanced condition.

The reason for this experiment is to ascertain the use of potentiometer as a strain gauge. It is expected that a same type of behavior will result when an actual strain gauge is attached to the model at any place of interest. Although this is hardwired, its wireless experiment is a projection of this concept of capturing analog current or voltage wirelessly to the computer. More on this is described in a separate Wireless Communication section.

So far, we have demonstrated distribution of current for balanced and unbalanced conditions using LED intensities. Now we use LabVIEW to the same application by noticing its response on the computer screen.

The **case** with potentiometer and LabVIEW demonstrated to the limit of breadboard circuitry by noticing the intensity of LED. We demonstrate this behavior on the LabVIEW screen using potentiometer as strain gauge.

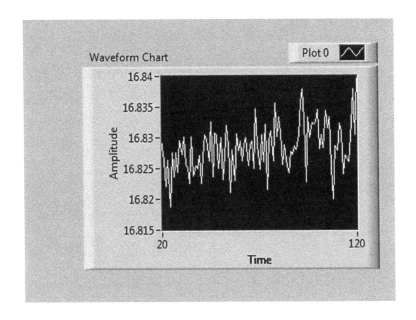

Figure 89. Initial screen.

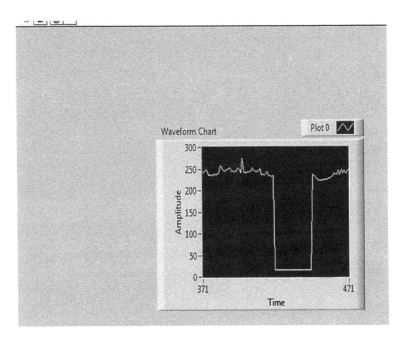

Figure 90. Response on LabView screen.

In figure 90, the response on LabVIEW screen to switch operation is noticed at upper line when the switch is closed.

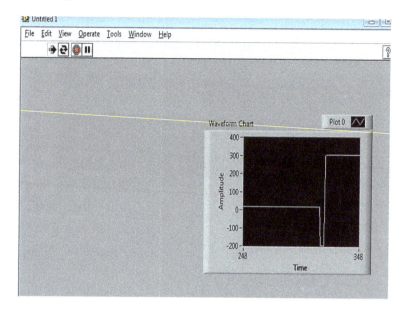

Figure 91. Response on LabVIEW screen to switch operation.

Figure 91 shows response on the LabVIEW screen when the switch is closed (displayed at upper line), with potentiometer engaged for balanced position. Change from initial condition at switch open followed by voltage changes progressively is noticed on the LabVIEW screen.

Figure 92 represents a balanced condition, with two LEDs showing same intensity of light. Another green LED in figure 92 is added for power input. The larger green LED is unlit at this stage.

In figure 93, the larger green LED shows current when the resistance is changed using potentiometer.

The lower line in figure 94 represents the case when resistance on potentiometer is changed for unbalanced condition. This is verified by the green LED in figure 93, when resistance is changed using potentiometer.

Figure 92. Switch is closed shows power via LEDs' balanced condition.

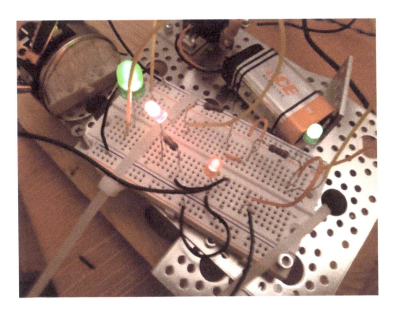

Figure 93. Reaction to change in resistance with potentiometer. The larger green LED is indicative of an unbalanced condition and activating LabVIEW amplifier.

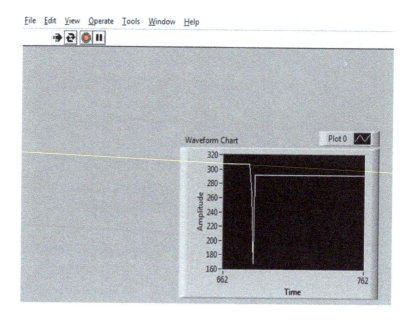

Figure 94. Upper line when power is up; upper lower line response to green LED light.

In figure 94, using the variable resister, the display shows unbalanced condition with amplifier circuit of LabVIEW engaged. For this case, LED with resister R2 shows a diminished intensity and the amplifier, and green LED is highlighted (figure 93). This example is typical behavior of strain gauge; however, it is simulated with potentiometer and connected to LabVIEW. The screenshot shown above without LabVIEW still applies to this system with LabVIEW circuit. LED behavior is identical.

Circuit Diagram for LabVIEW

These are very small voltages or current; thus, amplification is required to get any data.

Display of Devices

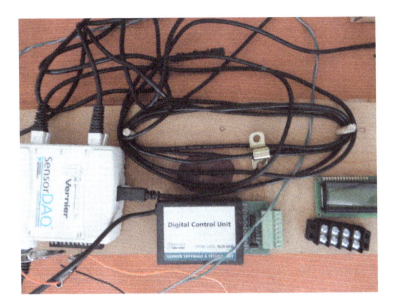

Figure 95. Image showing setup with Vernier. Courtesy of LabVIEW for Education.

Circuits for strain gauge, using these devices in the figure above are shown in appendix G.

Case 3 Aluminum Bar

This setup consists of three-wire strain gauge in quarter-bridge configuration. Material is an aluminum bar 1/8-inch thick, and 6.75 inches long, cantilevered from one end.

Test 1. Three-wire strain gauge and response with breadboard circuit using LEDs.

In real-world cases, use an aluminum bar (model on figure 96). A connection is made from the lead wires of strain gauge on aluminum bar to potentiometer terminal.

Test 2. Test the indicator at the amplifier with potentiometer by connecting the wire from the strain gauge to the proper terminals of potentiometer. We use a potentiometer to activate variable resistance on strain gauge. The behavior is exhibited on LabVIEW. Potentiometer is synonymous to strain gage.

Change the resistance by using the knob on potentiometer. The indicator at the amplifier will show a response by displaying the intensity (visible on green LED and on LabVIEW screen).

The test setup has a potentiometer used as a loading device. The three LED diodes represent the workings of the circuit. A green LED diode is inserted between the amplifier connection and at the point where the two resistors R3 and R4 join according to the circuit. The intensity of the green LED is seen to vary with the variation of resistance on the potentiometer. This is displayed on the LabVIEW screen (figure 102). Potentiometer is variable resistor similar to strain gauge and is used here to describe the analogy with strain gage. It is used here as a tool for varying the load; thus, introducing strain in the system. LabVIEW screen captures this behavior (figure 102). Figure 103 displays this change on breadboard and table 8 and 9 shows connections.

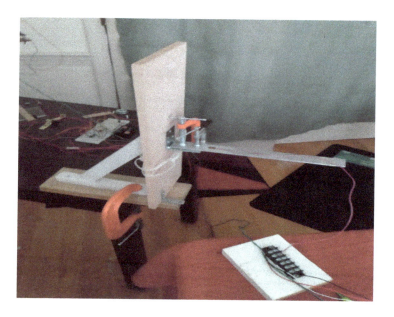

Figure 96. Demonstration on an aluminum bar with quarter-bridge strain gauge setup.

Connecter board

Strain gauge to connecter

Figure 97. Interface to LabVIEW sensor with connector board.

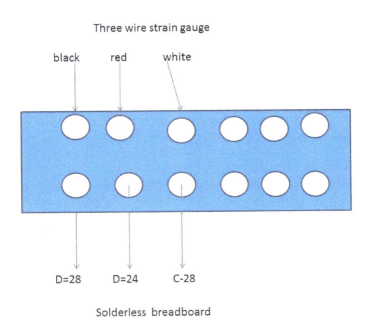

Figure 98. Showing lead lines on the connector board. D28, D24, and C28 are connecting points on the bread board.

Table 8. Three-wire setup from aluminum bar to connector board.

Description	Breadboard Column pin	Breadboard Row pin
Strain gauge r3 **Three wire setup** Potentiometer Connect white wire from strain gauge to c-28 node	c	28
Connect red wire from strain gauge to D-24 with R2 resistor	D	24
Connect black wire from strain gauge to amplifier node	d	28
Diode 1 @strain gauge Diodel anode	c	24
Diode 1 cathode	c	22

Table 9. Connection on breadboard.

Description	Breadboard Column pin	Breadboard Row pin
Strain gauge r3 **Three wire setup** Potentiometer Connect white wire from strain gauge to c-28 node	c	28
Connect red wire from strain gauge to D-24 with R2 resistor	D	24

Connect black wire from strain gauge to amplifier node	d	28
Diode 1 @strain gauge Diodel anode	c	24
Diode 1 cathode	c	22
Resistor r1	g	12
	i	18
Resistor r2	h	25
	g	29
Resistor r4	b	12
	b	16
Short red wire between R1 and r2	g-25	g-22
Diode d2 anode	F	12
cathode	D	12
Diode d3 @amplifier Cathode (green led)	i	14
anode	i	16
Amplifier ground Orange wire	F	14
Amplifier positive White wire	F	16
From Power supply positive to	J	25
From Power supply negative to	A	22

| Ground wire Black Short wire | B | 22 |
| Ground wire Black Short wire | A | 16 |

Figure 99. Image of breadboard model with switch open. Hardware for LabVIEW displayed with LED, potentiometer, and battery.

Image of connections of breadboard

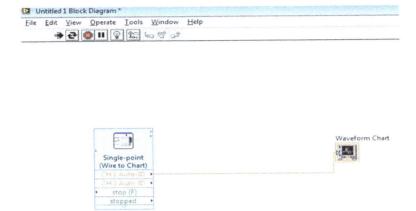

Figure 100. Computer screen display of connection from LabVIEW sensor to graphic display.

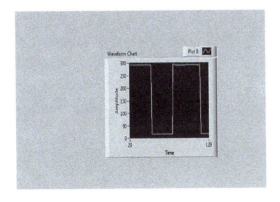

Figure 101. Response to change in potentiometer setting on LabVIEW.

Variation from change in potentiometer setting is shown on the display screen above.

Real cases on an aluminum bar

The model for real cases on an aluminum bar is shown in figure 96.

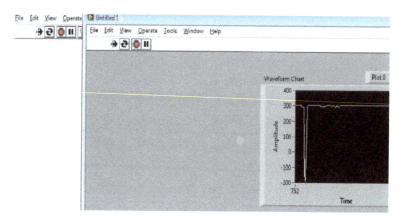

Figure 102. Response on LabVIEW from a change in potentiometer setting noticed at the upper line.

Case 1: aluminum bar

This change in potentiometer setting creates an unbalanced condition and its visual change is shown by varying intensities on three LEDs on breadboard (figure 103 below).

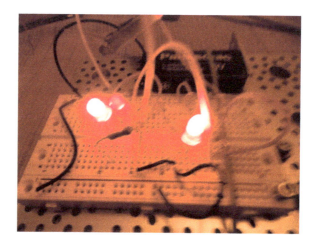

Figure 103. Showing the behavior on breadboard to the response above.

Case 2: plastic bar

This case tests the behavior by replacing potentiometer and using direct connection to the LabVIEW sensor. We apply variable loads to the sample. These are more flexible bars; thus, more prominent behavior is noticeable under variable loads. This model uses a two-wire strain gauge on a **plastic bar** of similar geometry as the aluminum bar. Demonstration with a plastic bar model is shown below with a quarter-bridge setup. Two leads are connected at the replacement points of potentiometer on breadboard.

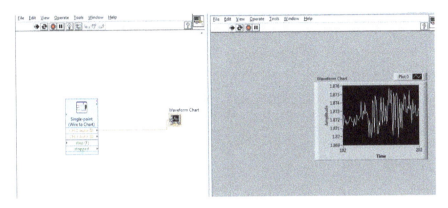

Figure 104. Initial view with switch open screen capture from LabVIEW.

Figure 105. Switch is closed (variation of voltage).

This behavior is with a plastic bar capturing strains under variable disturbance.

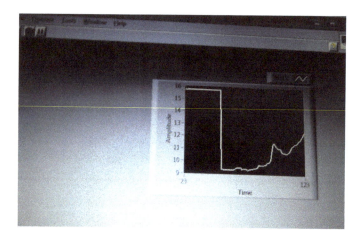

Figure 106. Response to applied disturbance with LabVIEW on plastic bar.

Case 3: wooden bar

This case uses a two-wire strain gauge with wooden bar.

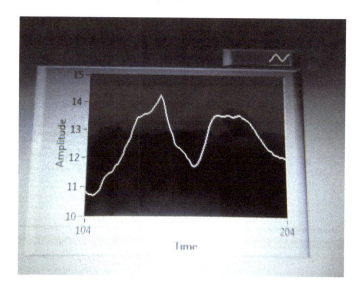

Figure 107. Two-wire strain gauge setup and response to small disturbance noted with LabVIEW.

Comparison of FEA and ESA on a Beam model

Figure 108. Setup with an aluminum bar model.

A 120-ohm strain gauge was affixed to the sample and tested for its function with **Micro-Measurements**[39] under wired condition. Its tabulation is noted in the table 11. The new feature by **wireless** testing is tested under the wireless communication using **Lord MicroStrain** setup.

The setup above (figure 108) has an aluminum bar and a strain gauge attached to the bar. This setup is with a quarter-bridge strain gauge. This is a three-wire setup. Two wires from one of the leads is used to allow for length adjustment. This setup is for room temperature.

---

39. From a class with Micro-Measurements of Vishay Precision Group.

Theoretical Stress Value

Theoretical stress and strain value computed using technical theory of beams is tabulated in table 10. This is for the model in figure 108.

Table 10. From technical theory of beams.

Method	Load	Moment (M)	SX		Stress value	Strain = Stress/ E strain	Computed strain
Theoretic	4	27 in lb.	0.0026		M/Sx = 10384 psi	10384/ 10600	0.9796226

Experimental Strain Value

Table 11. Data from quarter-bridge strain gauge setup on a cantilevered beam.

Method	output	setup			Stress E*Strain		Measured strain
Ex.Stress analysis from Micro, measurements	Quarter bridge strain gauge 4-4.10 lb. noted	Three wire			Stress= 0.999946*1060 0=10594.276		0.99946 strain
FEA model in next section							

When comparing the above values in strain with the two tables above, a difference in strain values is noticed. This could be due to surface preparation or imperfections in soldering. Above are typical sources that can lead to differences.

Wireless Communication

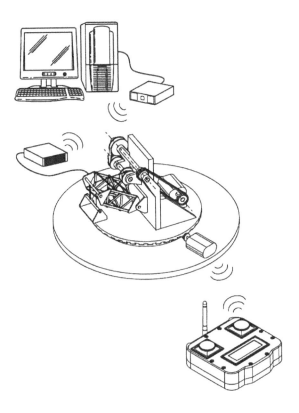

Figure 109. Wireless communication.

In this segment we address two features:

A run-time operation
Wireless communication and collection of data

Wireless communication uses sensor connect software with LORD MicroStrain devices. An example of a three-wire configuration of strain gauge is shown in figure 96. Table 12 shows the connection from connector board terminals to MicroStrain device.[40]

---

40. Connection to a transmitter device from LORD Micro Strain. (**With Permission**)

Table 12. Three-wire configuration.

Connector board terminals	Breadboard terminals	Micro strain terminals
Connect	From Figure 92	From user manual
Black wire from	D-28	To Sx-
White wire from	C-28	To Sx S
Red wire from	D-24	To GND (ground)

Finite Element Model on the Beam

We validate the accuracy of finite element program by comparing the results with the technical theory of beams on a simple cantilever bar (figure 108).

For this case, run a batch program **beams2** using the data parameters below:

6.75-inch-long beam (aluminum) E = 10,600,000psi
End load: 4 lb. at node 7
Moment of inertia in (in^4), d = thickness = 0.125, B = width = 1 inch
B (d^3)/12 = (0.125^3)/12 =0.001953125/12= 1.6276*10^ (-4)
=0.0001627 in^4
Section modulus = I/C = .0001627*2/(0.125) =0.0026032 inch^3
File name: Geom.dat
7 nodes
6 elements
Elasticity for aluminum
10600000.0
1
1 2
2
2 3
3
3 4

4
4 5
5
5 6
6
6 7
1 0.125 0.000167
2 0.125 0.000167
3 0.125 0.000167
4 0.125 0.000167
5 0.125 0.000167
6 0.125 0.000167
1 0.0
2 1.125
3 2.25
4 3.375
5 4.5
6 5.625
7 6.75
Force data
File name: Force.dat
1
7 4.0 0.0
Boundary condition at node 1, full fixity
File name: Bound.dat
1
1 1 1
Screen capture of results from finite element analysis:
1 force 4.0 lb.
Moment 27 lb. in
Length of bar = 6.75 inches
E = 10.6 *10^6 psi for aluminum

```
Command Prompt
0.29838434E+05  0.00000000E+00-0.14919217E+05  0.839205
0.12588089E+05 -0.83920596E+04  0.31470222E+04  0.000000
0.14919217E+05 -0.83920596E+04  0.00000000E+00  0.000000
0.62940444E+04  0.00000000E+00  0.00000000E+00  0.000000
     1       1       2
     1       1       2
 1 1 1
     1
0.00000000E+00  0.00000000E+00
     2
0.91159260E-02  0.15729442E-01
     3
0.34318782E-01  0.28598987E-01
     4
0.72391182E-01  0.38608626E-01
     5
0.12011573E+00  0.45758385E-01
     6
0.17427507E+00  0.50048221E-01
     7
0.23165178E+00  0.51478177E-01
     1  -0.40000558E+01
     2  -0.27000492E+02
```

Figure 110. FEA output on beam.

MicroStrain setup: strain data for 1 to 4 lb. load from MicroStrain wireless station is presented in figures 109 and 114, and table 12 and 13.

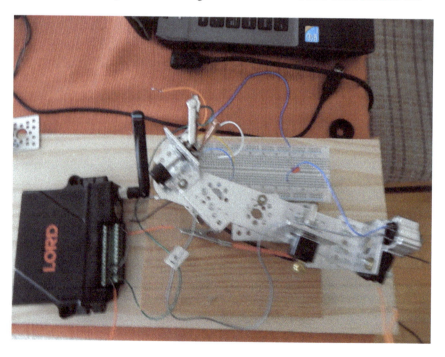

Figure 111. External hardware for MicroStrain device for controlling operation. (Lower left device from MicroStrain Inc.)

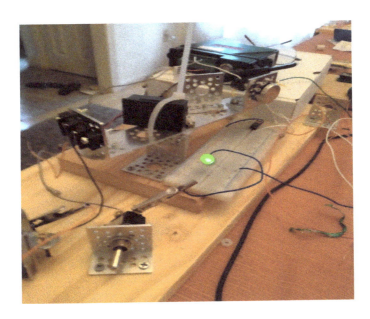

Figure 112. Green LED indicative of strain gauge engaged at switch close.

In figure 113, an alternative indicator replacing LED with an ammeter shows strain gauge is engaged.

Figure 113. On switch closed. Response is noted on an ammeter.

Strain for this setup was tabulated with **LORD MicroStrain station** with hardware shown above Data is included in the table 13. A plot is created using wireless data acquisition and is noted below. This is typically for the current strain gauge (with resistance of 120 ohm) for **callibaration of slope and offset** characterics.[41]

Table 13. Data from wireless devices.

Bits received at computer	Applied load in lb
128536	1
128553	2
128572	3
128607	4

In figure 114, a typical variation of load versus bits in a wireless setup, for slope and offset computation, is shown as well as a trend line.

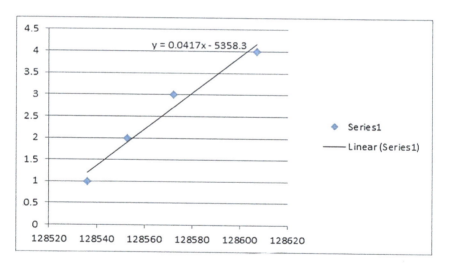

Figure 114. Slope and offset calibration.

41. LORD Sensing Systems. *LORD Sensing User Manual.* Vermont.page 67, 68 Used **with permission** from LORD Sensing System

Change in the bit versus load is shown on the plot. This is indicative of change in strain. Voltage (an analog value) is transformed into digital value (bits). This is the characteristic for this particular strain gauge. Using **Excel and scater function**, a bits versus force calibration equation is developed. Output equation for this strain gauge is posted on the graph above.

Caliberation result:

Slope = load/bits

Y (output) = slope*(bits) + offset

Table 14. Output values from slope and offset values applied to data of table 13.

Y (output) A	Slobe B	(*) Bits C	Product D = B*C	Offset From lot	Load (Y) Observed on Computer	Adjust offset	Adjusted offset & new offset	Rev Load Using new offset
Y1	0.0417	128536	5359.95	-5358.3	1.65	0.65	-5358.95	1
Y2	0.0417	128553	5360.66	-5358.3	2.36	0.65	-5358.95	1.71
Y3	0.0417	128572	5361.452	-5358.3	3.152	0.65	-5358.95	2.502
Y4	0.0417	128607	5362.912	-5358.3	4.6119	0.65	-5358.95	3.96

Comparing the difference between applied load and output using the equation it is noted that they are very close, but interference can make some difference. Otherwise the values are very close. The resulting strain from this experiment, FEA and technical theory of beams, indicate similar values.

When this strain gauge is affixed to the real model shown in the beginning of the manual, (Figure 1) to any member, a response matching finite element analysis is expected.

MicroStrain Diagrams

Block Diagram for Wireless Communication[42]

Block diagram in appendix F displays a circuitry for wireless devices. Readers may wish to refer to the manual in footnote 42 for details. Some excerpts from the manual are captured here. MicroStrain setup as shown in appendix F comprises of a few devices. These are transmitter, receiver, and computer. A transmitter receives signals (voltages) from a wired connection from strain gauge "leads" attached to the model (figure 109 and table 12). Transmitter transmits the signals of strain (indicating change in voltage) to the receiver wirelessly. Voltage signals (an analog value) are converted to bits (digital values). Receiver cable is connected to the computer. Computer runs the sensor connect software and displays the output graphically on the computer screen. Sensor connect software is used to calibrate the strain gauge by applying variable loads and collecting output into a database (table 13).

An Excel spreadsheet is used for plotting the collected data and developing a trend line. Calibrated equation posted on the trend line is input back into the sensor connect software with slope and offset (see footnote 42). At the switch operation shown above a continuous output of strain conditions using this equation is displayed on the computer for disturbances on the beam. Tables 13 and 14 summarize the above sequences.

Strain Gauge

Beam model (figure 108) is for a static case. The concept of beam model is projected onto the crane model. The crane model is operated with remote device for movement in the vertical plane.

The physical crane model is a maintained as nondestructive sample. The application of wireless action is demonstrated on a beam model that is more robust. The concept is projected to the crane model. Members stress levels can be monitored as the crane arm takes up different positions in the

---

42. LORD Sensing Systems. *LORD Sensing User Manual.* Page 47, I expanded this view with connection to devices and are shown in appendix F

vertical plane. These stress levels will be captured in a database from sensor connect software similar to the beam model, for examination. Collected values will allow comparison with the finite element analysis.

Using wireless devices, we calibrate the strain gauge for slope and offset by applying varying loads (see calibration curve, figure 114). We test the strain gauge on the beam model by applying varying loads and confirming the output received using sensor connect software on computer screen.

Force or load versus bit is calibrated from the strain gauge set up—in this case, with a quarter-bridge setup. Using the Excel software and the scatter of bits versus force, a calibration equation is developed. **Change in bits is indicative of changes in voltage.**

This equation is a pattern equation for strain versus force for this particular strain gauge. The equation allows examining the bit correspondence to force. In this case, it shows a strain induced on the gauge, corresponds to a 4-lb. load obtained from the equation. This allows for comparison with the theoretical finite element method on beam model. This indicates the same strain is observed using the strain gauge (experimentally) as that produced by finite element numerical method. This method allows for comparison of the two methods.

Figure 114 expresses this concept and use of the equation. Based upon the analog to bits received through the strain gauge from the wireless transmitter, this equation predicts a 4-lb. force was applied to the aluminum bar. If the same strain gauge is applied to any member of the truss, it is anticipated it will register the corresponding force in the member using this equation and running the wireless simulator. This can be compared with the force determined from the theoretical finite element model for different positions of the model arm as shown on animation page of model. It is expected to give matching results between FEA and real-time simulation of the crane model.

Summary of Comparison

Table 15. Tabulated values.

Analysis Type	End force at node 1	End moment at node 1	SX=1/C	adjustment	stress	Stress = E*strain	strain
Theoretic	4	27 in lb.	0.0026		M/Sx= 10384 psi		0.9796226
Exp. Stress analysis From "Micro, measurements"	Quarter bridge strain gauge, 3 wire setup						0.99946 strain Observed load 4.05+1b
FEA	3.999	27			10384psi		0.97962264
From Wireless[39] Calibration equation	Using the equation and load to bits relation	reflects a near 4lb load is applied to the systems at the end of bar.	Bits are conversion of volts sent from the transmitter	To the applied loads and reflect the strains in the gauge to this load		Thus it produces the same strain on the gauge corresponding to the loads.	

Conclusion

In this section we addressed two methods (1) LabVIEW as a wired connection and (2) MicroStrain device as a wireless method. We compared theoretical methods for stress/strain characteristics on a material with experimental values obtained from Micro-Measurements. We established a calibration equation for strain gauge with wireless device.

Using the calibration equation for this particular strain gauge and affixing to any member that is in motion, **it is expected to match up with theoretical results from FEA**. Results are dependent on calibration equation and surface preparation for strain gauge.

Appropriate account of elasticity will have to be taken into account when computing member-end forces with different material.

FEA results from a member-end force shown earlier can be compared with the force from the calibrated strain gauge when attached to the member on real model. A separate calibration on similar concept can be carried out using strain gauge on plastic material.

This introduces the concept of wireless technology at educational and practical level. This concept can be extended to real prototypes such as bridges or cranes.

PART 8:
Boundary Elements

Background

Boundary elements are other areas in solid mechanics.[43] These are alternative methods to finite element method typically for solution to two-dimensional problems (such as plane stress/plane strain cases).

The boundary element method uses the classical solution that is available and applies to a body of concern, typically using the boundary of the region. The classical solutions are an outcome from the theory of elasticity in mechanics. Unlike finite element that divides the region into smaller segments, this method subdivides the boundaries of the region (two sources where the applications are demonstrated are cited in the footnote 44). The image in figure 115 is symbolic of a cavity in a region enclosed with boundary elements. Such elements are subjected to internal or external disturbances.

---

43. S. L. Crouch and A. M. Starfield, *Boundary Element Methods in Solid Mechanics* (London, England: George Allen & Unwin, 1983), 25;
T.V. Hromadka II, and C. Lai, *The Complex Variable Boundary Element Method in Engineering Analysis* (New York: Springer Verlag, 1987). Pages 69,101

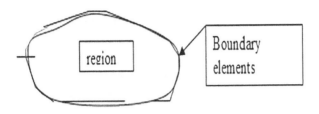

Figure 115. Boundaries subdivided.

The boundary element method uses subdivision of the boundaries enclosing a region as shown in the figure above. There are two approaches and they are distinct, one with real variable and another using complex variable. Applications to two-dimensional problems using these methods are described separately in the cited reference (see footnote 44) Interested reader may refer to these books for more details.

Conclusion

Boundary elements are alternate methods in numerical modeling to specific problems in two dimensions. Real and complex variable boundary elements are two distinct approaches in numerical modeling. These are cited in bibliography for interested readers for more information.

PART 9:
Data Translation or Data Exchange

Background

Another vast area is the framework for data translation or data exchange. It is a media of communication between participants to manufacture a product using a CAD model. In keeping with the objective of manufacturing, this integrates with a CAM tool and other areas as described in the figure of rapid prototyping. This is an application area of translational tools aiding in manufacturing of a product.

Data translation is a method of transforming the graphic's geometry to a text format so that it can be shared between different disciplines of interest. The importance of data translations lies in involving several working forces in production. Data translation is a media to connect all these working groups to achieve the final product. To perform a data translation, a geometric model called CAD model is needed.

Any geometry is made up of entities such as points, lines, curves, and surfaces. These fundamental entities are the building blocks of 3D objects. This introductory segment on data translation predominantly uses lines.

There are two popular methods. IGES-initial graphics exchange specification is one of the earlier standards. According to this standard, all types of curves are identified by a number. Similarly, all types of surfaces are given a number to identify the entity. Added to these IDs are their respective geometric attributes.

Standard for the Exchange of Product Model Data (STEP) is another standard established for transferring the CAD geometrical data. There are several protocols attached to this method. Some of these are AP203, AP214, AP227, and AP209. The method consists of transforming a mathematically defined entity to a text format. This is transferred to various agencies who in turn can invert the information and produce a model for their needs using their native algorithms. All geometrical figures are enclosed by boundaries as one unit called B-rep solid or boundary-represented solid.

In our approach, each individual piece of the model is transferred to a Gibbs CAM tool and fabricated. Complete assembly is manually assembled together.

Interested reader may view these methods of translation in their respective manuals cited in the bibliography.

Conclusion

This is a vast area—STEP with several specialized protocols, even more so. The manual for these are cited in bibliography as reference. These methods aid in transferring CAD data to computer-aided manufacturing machine facility, maintaining accuracy. First method (IGES) produces wire frame, and second method (STEP) provides B-rep solid.

PART 10:
Fabricated Model with CAM Tool

The model is an original architecture. Generated model sequences are shown below:

CAM

Application

Figure 116. CAM application.

Convert data for CAM tool

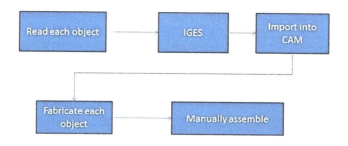

Figure 117. Typical flow path for CAM application.

A model is created from the 3D geometric data. Using data translation and Gibbs CAM machine, a model is fabricated. This model is displayed in figure 118.

Fabricated Model

A fabricated model from the outline created in graphics is shown below. Subsequently, another attachment was made. This model is further appended by attachment of motor, radio frequency control, and strain gauge. The crane model is kept as nondestructive sample, and strain gauges affixed on the model are conceptual and for display only. But the concept of a wireless method with strain gauge is applied to beam model. The concept is projected to real life on crane or bridges.

Figure 118. View of fabricated model.

Additional features added to the model: are controllers, belt-mounted pulleys, motor, chain, and remote operations (Figures 10 and 109). These aid in the mechanical operations of the CAM model.

For testing, the table(Figure 119) consists of a setup with aluminum bar for verifying experimental stress analysis and the model of figure 1 for performing remote operations.

Figure 119. Test table.

Model Analysis

Thus far, we have experimented with building a model. We presented different concepts using the model as backdrop. Projection to making prototypes and other real application usually requires model analysis. Thus, this concept can be projected to real field applications.

Appendix A

Mathematical Background

Reference is made to following topics in mathematics listed below:

Fourier series
Basic of theory of wavelets
Gram-Schmidt orthogonalization
Differential and integral calculus
Ordinary differential equations
Partial differential equations
Minimizing a function
Orthogonal vectors
Applied complex variable theories
Applied vector mechanics
Numerical analysis
Solving simultaneous equations
Eigenvalues, eigenvectors, problems, and associated frequencies
Coordinate geometry

Applied Computer Languages

Fortran, Java, C, and C++

Appendix B

Batch Programs

Stiffness Matrix

Beam. This is 4x4 matrix.[44]

See footnote 44 (page 200)

An alternative to that in the book cited in the footnote 44 is presented here.

In keeping with the view of being a survey , these methods are added.

Beam element comprises of 2 degrees of freedom at each node.

4x4 elements are populated using concepts of structural analysis such as using

Slope deflection method, moment area method or conjugate beam method.

The construct is as follows

For a beam constrained at both ends

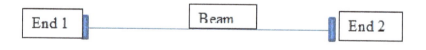

Figure 120. Beam with fixed ends.

---

44. Gere and Weaver Jr., *Analysis of Framed Structures1965 (Permissions noted)*
    McCormac, Jack C. *Structural Analysis*, Scranton, Pennsylvania, International
    Textbook Company, 1967 Page 398,399

1. A unit displacement given in downward direction of node 2 results in a moment 6EI/L^2 from integration method of differential equation.
   Similarly a moment is induced by deflecting the first node with respect to second node.
2. On a completely restrained beam a unit rotation at one end "node 1" causes a Moment of 4EI/L from slope and deflection equation. (See footnote 44, McCormac)
3. Using area moment the other end gets ½ this moment by equating the slopes to zero.
   When summed up for vertical equilibrium the resulting matrix is [S] in table 16
   L=length of member
   I= moment pf inertia
   E= elastic constant
   (With Permission, footnote 45, page 23, 47)

Table 16. Beam Stiffness coefficients.

	1	2	3	4
1	12	6L	-12	6L
2	6L	4L^2	-6L	2L^2
3	-12	-6L	12	-6L
4	6L	2L^2	-6L	4L^2

Corresponding stiffness matrix = [EI/L^3]*[S]*{u}I shown in (table 17)

Table 17. Beam Stiffness matrix.

Degree of freedom At nodes	1	2	3	4
1	12EI/L^3	6EI/L^2	-12EI/L^3	6EI/L^2
2	6EI/L^2	4EI/L	-6EI/L^2	2EI/L
3	-12EI/L^3	-6EI/L^2	12EI/L^3	-6EI/L^2
4	6EI/L^2	2EI/L	-6EI/L^2	4EI/L

Where "u" is a column of nodal displacements. Top row and left column show degree of freedom.

This is 4X4 matrix for beam element. The coefficients of table 16 are multiplied by EI/L^3

The same matrix can be derived by polynomial
Where a beam displacement polynomial is a cubic spline
w = a0 + a1 x + a2 x^2 + a3 x^3
Slope dw /dx = 0 + a1 + 2a2 x + 3 a3 x^2

w = trial function
X takes the values at 2 ends. End conditions are x =0 and x = L (table 18)

Table 18. Deflection and slope matrix.

[B] =

1	0	0	0
0	1	0	0
1	L	L^2	L^3
0	1	2L	3L^2

{a}

To eliminate the constant the matrix will be inverted to get it in the form of nodal displacement.
Matrix [B] can be inverted using Gauss Jordan method explicitly or numerically for larger matrix. Thus constants are expressed in terms inverted polynomial and nodal displacements.

{a} = [B]^-1 * {u}

Strain displacement is the next operation
Strain e = dw/dx = [0 1 2x 3x^2] {a}
Strain e = dw/dx = [0 1 2x 3x^2] *[B]^-1 * {u}

Where {a} is column vector of constants
Next a Stress/strain relationship is used and strain energy is computed.
Stress = E* strain
Strain energy is computed from the following expression

Strain energy = ½ integral [e(trans)] * E * [e]
Where e= strain and e(trans) its transpose
E = modulus of elasticity property

Appendix B

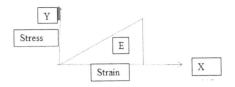

Figure 121. Stress – strain diagram.

From the diagram above strain energy is an area under the curve.

With appropriate substitution of strain in terms of nodal displacements and integrating the elements in the strain energy matrix , the resulting expression is shown below.

½ [k]{u}^2

Where [k] is the product of stress and strain.
Work done by external load equals [P][u}

The difference in Strain energy and the work done (Potential energy) leads to a Potential energy function PI as described on Figure 54 in the text.

PI =½ {u}trans[e(trans)]*E*[e]{u} – Pu

When the two expression are minimized for equilibrium the resulting equation is

[e(trans)] * E * [e] {u}= [P]

Or symbolically

[K]{u} = P

Where "u" is typical displacement.
trans = Transpose

Where the left hand side [K] is the stiffness matrix of beam same as above. Since this is lengthy only key steps are presented.

Final resulting matrix is as shown in tables 16-17

Plane Truss . This a 4x4 matrix shown below
Concept taken from footnote 44, page 231
Matrix of one direction displacement is shown below

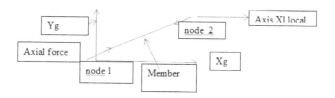

Figure 122. Axial force on a truss member.

For a typical member shown above the end forces induced at one end, while the other end constrained in axial direction results in a matrix [A] shown in (table 19)

From an expression Delta(deformation) = PL/AE, where delta is unit deformation from a unit load following matrix [A] is developed. The entries are multiplied by AE/L where

A = area of cross section of the member

E = member elasticity

L = length of member

Table 19. Resulting matrix for plane truss member.

$$[A] = \begin{array}{|c|c|c|c|} \hline 1 & 0 & -1 & 0 \\ \hline 0 & 0 & 0 & 0 \\ \hline -1 & 0 & 1 & 0 \\ \hline 0 & 0 & 0 & 0 \\ \hline \end{array}$$

This axial deformation is transformed in global axis of the structure, resulting in deformation in global X and Y direction.

DX DY are the direction cosines of the member local axis with respect to the global axis.

Assembled in matrix form this yields a 4x4 coordinate transformation matrix [C], table 20

Table 20. Table of direction cosines.

DX^2	DXDY	-DX^2	-DXDY
DYDX	DY^2	-DYDX	-DY^2
-DX^2	-DXDY	DX^2	DXDY
-DXDY	-DY^2	DXDY	DY^2

DX with respect to X axis

DY with respect to Y axis

Following computation below results in Stiffness matrix of plane truss (table 21).

Stiffness matrix = [C*trans)] * [A] * [C]

Elements are multiplied by a common multiplier AE/L

Table 21. Member Stiffnes matrix for Plane Truss.

DX^2	DXDY	-DX^2	-DXDY
DYDX	DY^2	-DYDX	-DY^2
-DX^2	-DXDY	DX^2	DXDY
-DXDY	-DY^2	DXDY	DY^2

Space truss. This is a 6x6 matrix for three degrees of system per node.

See footnote 44 (page 281)

Space truss member stiffness is similarly derived

This has three degree of freedom with each node. Globally. However the axial deformation is the same as the member in plane truss. The difference being to account for the third diction in Z direction.

The direction cosines are listed below

DX cosine of angle with respect to x axis

DY cosine of angle with respect to y axis

DZ cosine of angle with respect to z axis

Member stiffness matrix for this uniaxial case is shown below in [A] .

Space truss member has one directional displacement that is resolved in global direction

In terms of local member axis , the element deformation in axial direction of the member are shown in table 22. The matrix of table 22 is multiplied by a common multiplier AE/L

Table 22. Resulting matrix for space truss member.

[A] =

1	0	0	-1	0	0
0	0	0	0	0	0
0	0	0	0	0	0
-1	0	0	1	0	0
0	0	0	0	0	0
0	0	0	0	0	0

Direction cosines [C] for this case is shown in table 23

Table 23. Matrix of Direction cosines.

DX	DY	DZ	0	0	0
D2,1	D2,2	D2,3	0	0	0
D3,1	D3,2	D3,3	0	0	0
0	0	0	DX	DY	DZ
0	0	0	D2,1	D2,2	D2,3
0	0	0	D3,1	D3,2	D3,2

Multiplication of [C(trans)][A] [C] results in a 6x 6 matrix that is used in the program (table 24).

This matrix is in relation to global axis of structure.

Table 24. Global Stiffness matrix of space truss member.

DX^2	DXDY	DXDZ	-DX^2	-DXDY	-DXDZ
DXDY	DY^2	DYDZ	-DYDX	-DY^2	-DYDZ
DZDx	DZDY	DZ^2	-DXDZ	-DZDY	-DY^2
-DX^2	-DXDY	-DXDZ	DX^2	DXDY	DXDZ
-DYDX	-DY^2	-DYDZ	DYDX	DY^2	DYDZ
-DZDX	-DZDY	-DZ^2	DZDX	DZDY	DZ^2

A global transformation from a local axis to global axis is required to align the end forces applied in the global direction. This requires multiple rotations as cited in the reference (footnote 44, page 282).

Arbitrary orientation of a Member:
(Permission acquired on 3d figure of page 282)

(see footnote 44,page 282)

Transformation from member axis in arbitrary orientation to global axis is required.

This transformation is cited in the reference in 3D. Reader may refer to this page and derivation of rotation matrix. This 3D figure is represented in two dimensions in figures below.

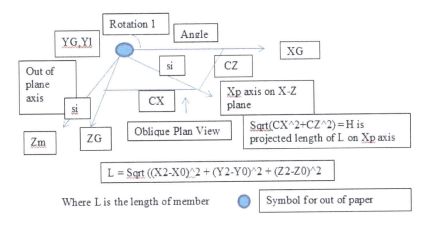

Figure 123. Plan View.

Following are with reference to above figure.

Transformation from global axis to the projected axis in the X- Z plane is defined by

Cos (si)= CX/H

Sin(si) = CZ/H

This transformation aligns local Zm with global Z axis.(from reference 44)

The directional attributes with respect to global axis are shown in table 25 (matrix [R1)

Table 25. Transform Global axis to projected axis.

Cos si	0	Sin si
0	1	0
-Sin si	0	Cos si

[Tm] = [R1[*[Tg]

where [Tm] = {Xp,Yl,Zl} and [Tg] = {Xg,Yg.Zg} column vectors arranged as row vectors.

Second alignment requires a rotation about Z axis relates the local Yl and H with respect to global X, and Y.(figure 124)

Looking towards Xp-YG plane (figure 124)

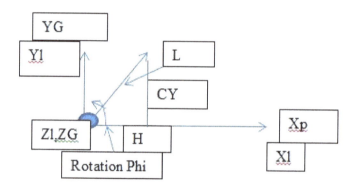

Figure 124. View on vertical plane.

Second rotation is on Zl by angle Phi.

This rotation gives additional relation between Yl and Xl local axis to the global axis.

Phi is an angle between Yl (local) and Yg global axis. (Elements of [R2] are shown in table 26).

[Tg} = [R2] * [Tm]

Where Tg and Tm are transformation column vectors, Tg = {Xg, Yg, Zg} and Tm = {Xp, YL, ZL}

Table 26. Transformation matrix for figure 130.

[R2] =

Cos phi	Sin phi	0
-Sin phi	Cos phi	0
0	0	1

Where Sin (phi) = CY, Cos (phi) = H

Final transformation [RT] is a product of [R1][R2][R3]

[R3] is rotation matrix for aligning principle axis of member with global axis. These effect only flexural elements of the stiffness matrix.

Space frame. This is a 12x12 matrix with six degrees of freedom system. See footnote 44 (pages 198, 282, and 284)

This element is assigned 6 degrees of freedom at each node , three displacements and three rotations. Sequence of displacements is followed with rotations at node. Displacement in the positive global direction is taken as positive. Counterclockwise rotation is taken as positive using the right hand rule convention.

The displacement and rotation matrix in X and Z direction are similar to beam elements except that the placed with reference to global degree of freedom in the matrix. Values are identical.

This will account of 4 degrees of freedom.

One degree of freedom is assigned the axial deformation similar to space truss.

6 th degree of freedom is assigned to a twist term about member axis.

$G*[Ix]$, where G is a shear modulus, $Ix = Sum(1/3*(b)*t^3)$.

Where b = width of element

T = thickness of that element

Ix is the torsional rigidity, and is made up of width and thickness of individual element of an open section, such as an I section., or polar moment of inertia of closed sections.

This completes the assigned values at one node. Since the matrix is symmetric the same values are applied to the second node.

Coordinate transformation is required for general placement in global system. Similar to above.

There is one additional rotation that is required to orient the principal axis of members in global position[R3].

This can be referred to in the cited reference.(page 291)

Structure of 12x12 Matrix is presented here. 2 quadrants are presented in tables below.

The lower left quadrant is mirror of upper right quadrant. And lower right quadrant is same as Upper left quadrant. Reader can refer to the cited reference pages for more information.

Final global stiffness matrix is a product of

[RT]trans][S][RT]

Where [RT] = [R1][R2][R3]

Where [RT] is applied to 2 end nodes of member. [S] is a full 12x12 matrix developed from partial matrix shown in tables 27 and 28

Table 27. upper left of matrix at 1.

Degree Of freedom	1	2	3	4	5	6
1	delta x-1	delta y-1	delta z -1	rot x-1	rot y-1	rot z-1
1	A(1,1)	0	0	0	0	0
2	0	S(2,2)=12EI/L^3 I=Iz	0	0	0	S(2,6)=6EI/L^2 I=Iz
3	0	0	S(3,3)=12EI/L^3 ,I=Iy	0	S(3,5)= -6EI/L^2 I=Iy	0
4	0	0	0	S(4,4)=G*I/L I=Ix	0	0
5	0	0	S(5,3)= -6EI/L^2 I=Iy	0	S(5,5)=4EI/L I = Iy	0
6	0	S(6,2)=6EI/L^2 I=Iz	0	0	0	S(6,6)=4EI/L I=Iz

Table 28. Upper right of matrix at 2.

Degree Of freedom	7	8	9	10	11	12
	delta x-2	delta y-2	delta z -2	rot x-2	rot y-2	rot z-2
1	A(1,7=-A(1,1)	0	0	0	0	0
2	0	S(2,8)= -12EI/L^3 I = Iz	0	0	0	S(2,12)= 6EI/I^2 I = Iz
3	0	0	S(3,9)= -12EI/L^3 I= Iy	0	S(3,11)= -6EI/L^2 I = Iy	0
4	0	0	0	-GI/L I=Ix	0	0
5	0	0	S(5,9)=6EI/L^2 I=Iy	0	S(5,11)= 2EI/L I=Iy	0
6	0	S(6,8)= -6EI/L^2 I =Iz	0	0	0	S(6,12)=2EI/I I=Iz

Final stiffness matrix is

[RG]tans * [S]* [RG]

[RT] is a matrix of 3x3. [RG] is a 4x4 matrix with 4 blocks of-3x3-[RT] matrix placed along the diagonals and off diagonals being 0. Each block will have one 3x3 [Rt] matrix. Complete [RG] matrix will be 12x12. The final computation puts the 12x12 stiffness matrix in the global position. The forces are applied in the global directions. Total stiffness matrix is contribution of all the elements as shown in the text.

Plate in plane stress & strain.[45]

See footnote 45 (pages 69, 88, 89)

Software uses the content from the cited pages.

(With permission)

Plate in bending.

See footnote 45 (pages 110 and 111)

Software uses the content from the cited pages.

---

45. K. C. Rockey, et al., *The Finite Element Method: A Basic Introduction for Engineers.* New York: John Wiley & Sons, 1983. Permissions noted

(With permission)

The following examples are some of the batch programs that can be executed from the thumb drive of Assembly class:

1 for Beam analysis; Beam.bat

    For static analysis Beams.bat

    For eigenvalue extraction Beamd.bat

    For space truss: spaceTruss.bat (this file is for simulation)

    For frame analysis: mast.bat

    Other batch programs are listed in software package.

Typical Structure of Batch files

Batch file	Executable files
Beam.bat	Beam2
	Load
	Disp
	Bound
	Bsolvel
	Resultant
	Beamfrce
	Dyanl2
BeamS.bat	Static analysis
beamD.bat	Dynamic analysis
Sptruss2.bat	
	Sptruss2
	Load
	Disp
	Bound
	Bsolvel
Plbendr.bat	Rectangular elements
Plbendt.bat	Triangular elements

Executable Programs

Steps in activating Fortran programs:

For finite element program

Copy of these executables are saved on thumb drive labeled Fortran.

Listed are Fortran executables that can be activated from java GUI.

This path is predefined in Java Assembly class.

For information:

Select a USB port, label this as drive F:>
Insert the second thumb drive (labeled Fortran.exe)
Create following four folders. Only if one wants to compile. Otherwise not necessary
They exist on the thumb derive with Fortran executables.
Move to the folder Fortrancom
Move to the folder Fort99
Move to the folder G77

The <u>following is the batch file called CCFOR</u> for creating executable files.

Table shows typical Fortran program files compiled to make executables.

1. Function name	Argument	switch	executables
c>g77	LOAD.FOR	-o	Load.exe
c>g77	DISP.for	-o	Disp.exe
c>g77	UDL.FOR	-o	Udl.exe
c>g77	BEAM2.FOR	-o	Beam2.exe
c>g77	BOUND.FOR	-o	Bound.exe
c>g77	SPTRUS2.FOR	-o	Sptrus2.exe
c>g77	RDLOAD.FOR	-o	Rdload.exe
c>g77	BSOLVE1.FOR	-o	Bsolve1.exe
c>g77	Dyanl2_1.FOR	-o	Dyanl2_1.exe
c>g77	Dyanl3	-o	Dyanl3.exe
c>g77	Plbendr	-o	Plbendr.exe

Related Input files/output files

Generate Triangular mesh: requires input of number of rows and column ,it is saved automatically to a file..

Generate Rectangular mesh: requires input of number of rows and column ,it is saved automatically to a file..
Image is Viewable on Diagram.java

Beams (static cases) (source file Beam2.for)
Batch file: Beams.bat

beam2.exe (develops stiffness matrix) Input files-geom.dat(Geometry of beam)

sprngk (adds spring constants) (input file- spring.dat)

load (develops load vector) input file -(force.dat) ,output file- fload.dat

udl (develops uniformly distributed load vector) Input file- udl.dat,internal file- geom.dat ,

output file- fload.dat

disp (applies displacement to input) input file- dis.dat, output file- fload.dat

rdload2 (reform load vector based on boundary conditions) input file developed internally

bound (applies boundary conditions) input files- bound.dat

output files- sptrstbn.dat, fload.dat

bsolve1 (solves the matrix) internally developed input file- sptrdt, sptrstbn.dat, bound1.dat, fload.dat, Output files- w.dat,

Resultant (computes resultant at boundaries) internally developed Input files:(sptrstf.dat , w.dat, sptrdt,bound.dat) , Output file- resultant.dat,

elreac (results with elastic spring constants)

(internally developed Input files- sptrdt, sptrstf.dat, w.dat)

output files (elreac.dat)

beamfrce (computes beam end forces),Internally developed Input files: (geom.dat, w.dat,),

Output files- beamfrce.dat

Plane Truss
Batch file :ptruss.bat

Main source file PlTruss.for (input files :Pltruss.dat, ncoord.dat, icase.dat)

(Icase.dat developed internally for even and odd case for conceptual simulation)

Load (Input file - force2.dat)

Bound Input file -BOUNDPTruss.dat

bsolve1 (output file - w.dat) for displacements.

resultant (output file -:resultant.dat)

ptrussfrce (output file - ptrussfrce.dat)

mendtrussfrce(output file overwrites -ptrussEndfrce.dat)

Plane Frame(pframe.bat)

(PlFrame.FOR) (input file-PlFrame.dat)

Load(input file-force4.dat)

Bound (Input file- bound4.dat)

bsolve1(output file: -w.dat)

resultant (output file - resultant.dat)

pframefrce(output file overwrites - plframe.dat)

mendframefrce(output file- memendfrce1.dat)

Grid Batch file: (grid.bat)

PlGrid.for (Input file -PlGrid.dat)

Load(input file- force5.dat)

Bound Input file -(boundPlgrid.dat)

bsolve1(output file - w.dat)

Space truss (Sptrus2.for, D:/fnprj/SPTRUSS.txt sptruss2.bat)

PlTruss.for (generates planer truss coordinates at run time)

GenDataSpTruss (generates space truss joint coordinates)

GenNewSptruss (reads data from GenDataSpTruss and saves as SPTRUSS.txt) sptrus2 (input file- D:/fnprj/SPTRUSS.txt), Output file- sptrstf.dat

Load (input file-force3.dat)

Bound (input file - boundSptruss.dat)

bsolve1(output file -w.dat)

resultant (output file -resultant.dat)

sptrussfrce (output file overwrites- sptrussfrce.dat)

sptmemendfrce (output file overwrites: sptendfrce.dat)

Space Frame (mast.bat) for displacements only

(Spframe.bat) for resultants

Source file and Input/output files:

mast1.for (Input file -mast1.dat)

frmld1(input file - mastFORCE.dat)

bound(input file- frmBOUND.dat)

bsolve1 (output file- w.dat)

resultant (output file-resultant.dat)

frameGLendfrce (global end forces overwrites : spfrmfrce.dat))

**Plate in Plane stress triangular element
plstressT.bat**

Tripl (input file - nodedataT.dat)

Load (input file - force8.dat)

Bound(boundTPSTRS.dat)

bsolve1(output file- w.dat)

triplnstress(output file- triplstress.dat)

**Plate in Plane Stress triangular and truss element
See Appendix E
Tripltruss.bat**

Input file -(nodedataT.dat)(element type :elType.DAT)(plateProp.DAT)

load (input file - force11.dat

bound(input file - boundPL11.dat)

bsolve1(output file- w.dat)

resultnt (output file- resultant.dat)

Plate in Plane stress rectangular element
Batch fle-plstressR.bat

Rectpl (input file-platebnd.dat)

load (input file- force7.dat)

bound(input file -boundPL9.dat)

bsolve1(output file-w.dat)

rctplst (out put file- rctplnstrs.dat)

Plate in bending with triangular element
Batch file-Plbndt.bat

plbendT1(input file- nodedataT.dat)

load(input file-force10.dat)

bound(input file- boundPL10.dat)

bsolve1(output fil-:w.dat)

tplstrs (output file- tplstrs.dat)

Plate in bending with rectangular element (plbendstrs.for)
Plbnd1.bat

PLBENDR.for (Input file: nodedataR.dat, plateprop.dat)
load(Input file -force9.dat)

bound(Input file- boundPL9.dat)

bsolve1(Output file-w.dat)

plbendStrsR(output file- plbstress.dat)

Orthogonal vectors
BeamsOr.bat

beam2 (geom.dat) (input same as beam)

load (input file - force.DAT)

udl

rdload2 (input files :internal)

bound input files: bound.dat)

bsolve2(Output file: (orthogonal vectors in ritz.vec)

File for Dynamic analysis
Beamd.bat (batch file)

beam2(input file: geom.dat)

load

disp

rdload2

bound

cndnse

matinv

dyanl2_1(eigen values and vectors in Out.vec)

Other programs developed by writer are "Ginv.for"(for matrix inversion by Gauss Jordan method) and

"Gsolve2.for" (Gaussian elimination on fully populated matrix)

Appendix C

Computer Graphics

The image below shows the steps in activating graphics programs using CAD/CAE/CAM option button. Code reuse is common in software engineering. This is an example of code reuse where the original software was for a different requirement (from writer's class notes). Applicable portions have been extracted and tailored for this application. This is only conceptual.

How to run CAD CAM CAE

Batch file

Type - Cadcam

Java and Fortran (g77 compiler are to be residing on the computer)
At this writing both are available as free downloads.(Fort99 for Fortran)
Requires two thumb drives, one for Java another for Fortran program

To run the program, user must run the following initialization:
Go to c :\users\Admin from this drive, must type "setup2"
This action will invoke a batch file setup2 .bat with selected path.
A batch file would need to be created such as setup2(Example below)
in the root drive C. :\users\Admin

The following listed items must be residing in the path and class path, in the batch file, as shown in the image below.

For Fortran :program Fort99. will be unzipped to a folder C:\f. a batch file

g77setup will need to be activated for Fortran compiler.(as shown in the documentation of FORT99 in g77 folder)

Setup2 Example:

set path= c:\Program Files (x86)\Java\jre1.8.0_171\bin; c:\Program Files (x86)\Java\jdk1.8.0_05\bin; d:\fnprj\org\gjt\mm\mysql; c:\mysql\bin;c:\fnprj\xerces-1_2_0**;%path%**

set CLASSPATH=.;d:\fnprj\cscie247\asn6;d:\fnprj\org\gjt\mm\mysql; c:\Program Files (x86)\Java\jdk1.8.0_05\bin; c:\Program Files (x86)\ Java\jre1.8.0_171\lib;c:\mysql\bin;d:\fnprj\dom;c:\fnprj\xerces.jar;c:\ fnprj\Xerces-J-src.2.12.0.zip;c:\fnprj\xalan-j_2_3_1-bin.zip;c:\fnprj\ xerces-1_2_0\samples\dom;c:\fnprj\xercesSamples.jar;**%path%**

This setup2 is a batch file, typical for this application. All the folders and path reside on this drive. This file sets up path and class path in the root, allowing the executables from any drive.

Typical to this setup are java, jdk, xerces. Similar path is created for Fortran compiler in "\F\G77" folder .

```
set path=C:\Program Files (x86)\Java\jre1.8.0_431\bin;c:\Java-app\fnprj\org\omg
\corba-api-5.0.0.jar;c:\Program Files(x86)\Java\jre1.8.0_431\bin;c:\mysql\bin;C:
\Program Files (x86)\OpenLogic\jdk-8.0.432.06-hotspot\bin;d:\java\j2re1.4.0_01;
%path%
set CLASSPATH=.;C:\Java-app\fnprj\org\omg\corba-api-5.0.0.jar;c:\Program Files
(x86)\OpenLogic\jdk-8.0.432.06-hotspot\bin;d:\fnprj\cscie247\asn6;c:\fnprj\Java
\jdk1.7.0_79\lib;d:\fnprj\xerces-1_2_0;d:\fnprj\xerces-1_2_0\samples\dom;%path%
```

Next change to D: drive and CD to " fnprj " directory.

Use the thumb drive labeled with Java Programs in black thumb drive and insert into a USB port. Label this as D drive and CD from this drive to "fnprj".

The code is designed with two options to get client display screen:

Option 1: type "Assembly" at prompt from D drive (A batch file Assembly.bat is invoked) **finish** button on each traverses to next program

Option 2: This example uses a concept from distributed objects and uses client-server application:

- Type "CADCAM" at prompt (CORBA session is started, CADCAM is a batch file)
- Invoke **tnameserv** (transient naming service) , the path is in "Program Files\Java\jdk\bin"
- From distributed objects, a sequence of prompt is displayed from this batch program, requiring a return action after each prompt
- Next PlaneServer text is displayed (a return at prompt starts a server)
- Following this step a plane server page is displayed
- Next a PlaneServerDriver text is displayed <return>
- A "client" prompt is displayed (for first client)<return>
- A "client" prompt is displayed (for second client))<return>
- Resulting action will display program option screen
- Select CADCAM from lower left button in green]
- A button press turns the color to "red" indicating notification
- Next CAD/CAE/CAM workflow screen is displayed
- Other buttons are inactive and left for future additions.
- Selecting "Start" will invoke "Assembly.class" GUI shown in the manual
- Other programs are listed and executed according to the description in the manual.

Appendix D:
Circuit Diagrams

This appendix shows a circuit diagram using breadboard and devices.

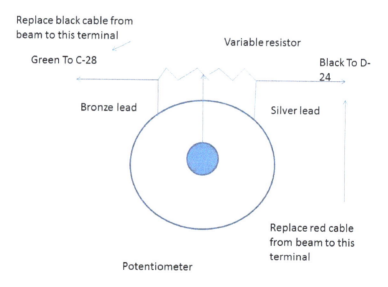

Figure 125. Potentiometer and connecting terminals with color-coded cables.

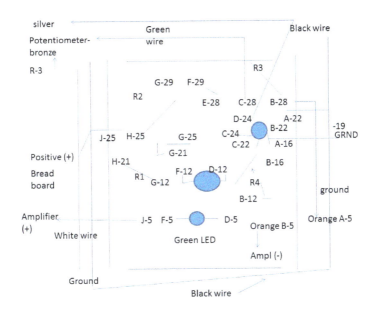

Figure 126. Breadboard nodes with color-coded wires.

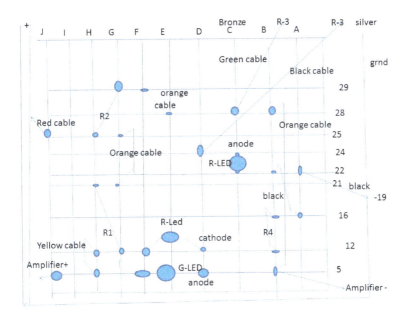

Figure 127. Breadboard grid with LEDs.

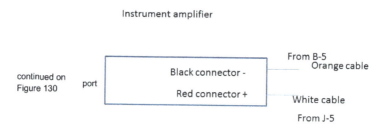

Figure 128. Connection at amplifier terminals.

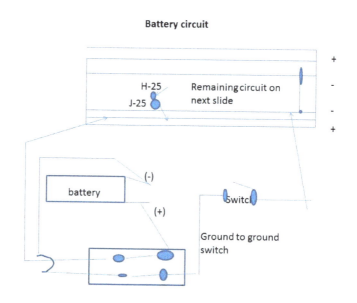

Figure 129. Connection to battery, breadboard, and operating switch.

Appendix D: Circuit Diagrams

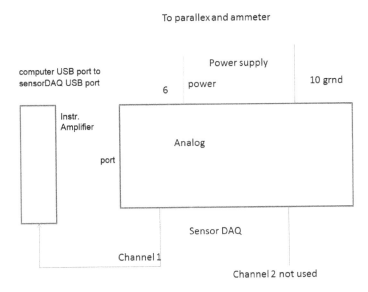

Figure 130. Analog sensor terminals to computer and external power source.

Appendix E

Plane Truss and Plate element

Below is a data structure for the finite element model, a combination of truss and continuum. Each node has two degrees of freedom (for displacements in x and y direction).

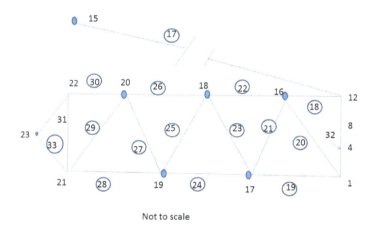

Figure 131. Truss part of completed model.

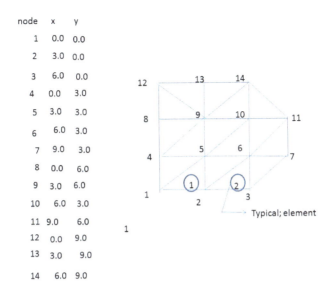

node	x	y
1	0.0	0.0
2	3.0	0.0
3	6.0	0.0
4	0.0	3.0
5	3.0	3.0
6	6.0	3.0
7	9.0	3.0
8	0.0	6.0
9	3.0	6.0
10	6.0	3.0
11	9.0	6.0
12	0.0	9.0
13	3.0	9.0
14	6.0	9.0

Figure 132. Right side attachment and plane stress FE model.

23 (total number of nodes)
33 (number of elements)
1 0.5 (thickness)
2 0.5
3 0.5
4 0.5
5 0.5
6 0.5
7 0.5
8 0.5
9 0.5
10 0.5
11 0.5
12 0.5
13 0.5
14 0.5
15 0.5

16 0.5
17 10.0 (area)
18 10.0
19 10.0
20 10.0
21 10.0
22 10.0
23 10.0
24 10.0
25 10.0
26 10.0
27 10.0
28 10.0
29 10.0
30 10.0
31 10.0
32 10.0
33 0.5
1 1 2 5
2 2 3 6
3 3 7 6
4 4 5 9
5 5 6 10
6 6 7 11
7 12 8 9
8 9 10 14
9 10 11 14
10 1 5 4
11 2 6 5
12 4 9 8
13 5 10 9
14 6 11 10
15 12 9 13
16 9 14 13
17 12 15 15

18 12 16 16
19 1 17 17
20 1 16 16
21 16 17 17
22 16 18 18
23 17 18 18
24 17 19 19
25 18 19 19
26 18 20 20
27 19 20 20
28 19 21 21
29 20 21 21
30 20 22 22
31 21 22 22
32 1 12 12
33 22 23 21 last element approximated as one triangular element
    1 0.0 0.0 4 1
    2 3.0 0.0 4 1
    3 6.0 0.0 4 1
    4 0.0 3.0 4 1
    5 3.0 3.0 4 1
    6 6.0 3.0 4 1
    7 9.0 3.0 4 1
    8 0.0 6.0 4 1
    9 3.0 6.0 4 1
    10 6.0 6.0 4 1
    11 9.0 6.0 4 1
    12 0.0 9.0 4 1
    13 3.0 9.0 4 1
    14 6.0 9.0 4 1
    15 -42.0 29.0 4 1
    16 -7.0 9.0 4 1
    17 -14.0 0.0 4 1
    18 -21.0 9.0 4 1
    19 -28.0 0.0 4 1

20 -35.0 9.0 4 1
21 -42.0 0.0 4 1
22 -42.0 9.0 4 1
23 -50.0 4.5 4 1

Boundary conditions
From boundpl11.dat file
2
15 1 1 restrained in x and y direction
23 1 1

From force8.dat file
forces
2
7 0.0 0.5
11 0.0 0.5

Data-defining element type, one - plate element, two- planar truss element
1 1
2 1
3 1
4 1
5 1
6 1
7 1
8 1
9 1
10 1
11 1
12 1
13 1
14 1
15 1
16 1
17 2

18 2

19 2

20 2

21 2

22 2

23 2

24 2

25 2

26 2

27 2

28 2

29 2

30 2

31 2

32 2

33 1

Displacement result at node and ux, uy

1

    0.51894941E-03 0.14357838E-02
    2
    0.61674061E-03 0.16380395E-02
    3
    0.66517218E-03 0.18564161E-02
    4
    0.45556901E-03 0.14706713E-02
    5
    0.46334264E-03 0.16260804E-02
    6
    0.48188149E-03 0.18551303E-02
    7
    0.49636373E-03 0.21316286E-02
    8
    0.32758861E-03 0.14698218E-02
    9
    0.32086656E-03 0.16254999E-02

10

0.30463858E-03 0.18538409E-02

11

0.29208671E-03 0.21253889E-02

12

0.26738501E-03 0.14227822E-02

13

0.16983874E-03 0.16416542E-02

14

0.11818334E-03 0.18472133E-02

15

0.00000000E+00 0.00000000E+00

16

0.24121553E-03 0.12192908E-02

17

0.47038839E-03 0.10405673E-02

18

0.18944539E-03 0.82157820E-03

19

0.42125874E-03 0.64080104E-03

20

0.13824407E-03 0.42020064E-03

21

0.37156034E-03 0.23825456E-03

22

0.11292775E-03 0.22617861E-03

23

0.00000000E+00 0.00000000E+00

Reactions at the supports

```
29  0.20841672E+01
30  -0.99246067E+00
45  -0.20841763E+01
46  -0.75720651E-02

:\FortranCom\FortranCom\Fort99\G77>_
```

Figure 133. Screen capture of computer output.

Reactions at the support point 15 and 23 are shown in the image; 29 is indicative of horizontal force and 30 is indicative of vertical force at support point 15. Similarly, 45 is the degree of freedom at support point 23 and indicative of horizontal force; 46 is the degree of freedom associated with vertical reaction at support point 23. Vertical reactions are balanced between two support points.

Appendix F

Circuit Diagrams

Wireless communication

From Micro Strain device

Micro Strain Devise configuration

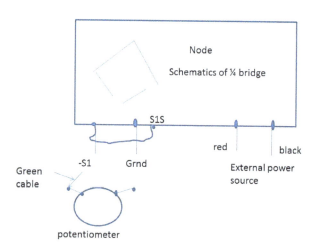

Figure 134. Connection to MicroStrain transmitter device terminals.

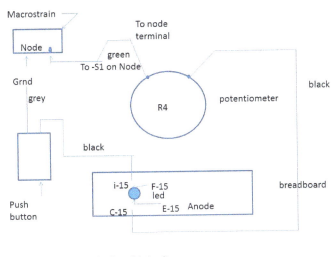

Resistor R4 circuit

Figure 135. Potentiometer for R4 resister and to transmitter (node).

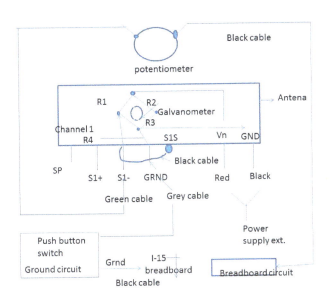

**Figure 136. Potentiometer to transmitter node with labeled cables.
(Galvanometer symbolic node of internal bridge circuit.)**

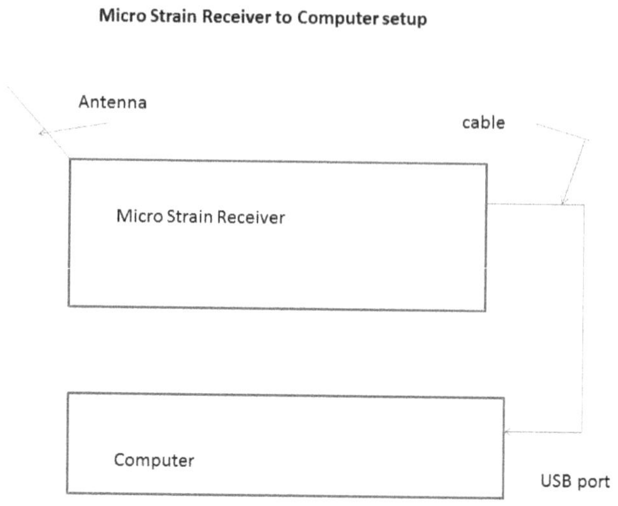

Micro Strain Receiver to Computer setup

Antenna

cable

Micro Strain Receiver

Computer

USB port

Figure 137. MicroStrain receiver to computer.

Appendix G: Image of Bridge

View

Figure 138. Possible field application on a bridge.

Some videos showing the operations on the model are saved on a thumb drive.

Bibliography

Angell, Ian O. and Gareth Griffith. *High-Resolution Graphics Using FORTRAN 77*. New York: John Wiley & Sons, 1987.

Atkinson, Kendall E. *An Introduction to Numerical Analysis*. New York: John Wiley & Sons, 1978.

"Application of strain gauges in measuring strain." Rayleigh, NC: Micro Measurements Co. Vishay Measurements Group, INC

Boggess, Albert and Francis J. Narcowich. *First Course in Wavelets with Fourier Analysis*. London, England: Pearson, 2001.

Biggs, John M. *Introduction to Structural Dynamics*. New York: McGraw-Hill, 1964.

Carnahan Brice, H. A. Luther, and James O. Wilkes. *Applied Numerical Methods*. New York: John Wiley & Sons, 1969.

Crouch, S. L. and A. M. Starfield. *Boundary Element Methods in Solid Mechanics*. London, England: George Allen & Unwin, 1983.

Dally, James W. and William F. Riley. *Experimental Stress Analysis*. New York: McGraw-Hill, 1978.

Date, C. J. *An Introduction to Database Systems*. Boston: Addison-Wesley, 1982.

D'Azzo, John J. and Constantine H. Houpis. *Linear Control System Analysis and Design*. McGraw-Hill, 1973.

Desai, Chandrakant S. and John F. Abel, *Introduction to Finite Element Method: A Numerical Method for Engineering Analysis*. New York: Van Nostrand Reinhold Co., 1971.

Divakaran, S. and V. K. Garg. *Strength of Materials*. London: Asia Publishing House, 1969.

Dym, Clive L. and Irving H. Shames. *Solid Mechanics*. New York: McGraw-Hill, 1973.

Farley, Jim. "Java for Distributed Computing", O'REILLY & Associates, Sebastopol, CA,1998.

Frocht, Max M. *Photoelasticity*. New York: John Wiley & Sons, 1941.

Gerald, Curtis E. *Applied Numerical Analysis*. Boston: Addison-Wesley, 1978.

Gischner, Burton. *Ship Common Information Model (SCIM)*. Newport News, VA: National Shipbuilding Research Project All Panel Meeting, 2012.

Groover, Mikell R. and Emory W. Zimmers, Jr. *CAD/CAM: Computer-Aided Design and Manufacturing*. Hoboken, New Jersey: Prentice Hall Inc., 1984.

Hayes, Thomas with Paul Horowitz. *Learning the Art of Electronics: A Hands-On Lab Course*. Cambridge: Cambridge University Press, 2016.

Hessop, H. T. and F. C. Harris. *Photoelasticity: Principles and Methods*. New York: Dover Publications Inc., 1949.

Higdon, A., E. H. Ohlsen, W. B. Stiles, and J. Weese, eds. *Mechanics of Materials*. New York: John Wiley & Sons, 1978.

Horstmann, C. S. and Gary Cornell. Java Series. Sun Microsystems Press.

Hromadka II, T.V. and C. Lai. *The Complex Variable Boundary Element Method in Engineering Analysis*. New York: Springer-Verlag, 1987

Initial Graphics Exchange Specification (IGES), .U.S. National Bureau of Standards, NBSIR 80-1978.

Irons, B. and N. Shrive. *Finite Element Primer.* New York: John Wiley & Sons, 1983.

Jenkins, W. M. *Matrix and Digital Computer Methods in Structural Analysis.* New York: McGraw-Hill, 1969.

Kresysig, Erwin. Advanced Engineering Mathematics. New York: John Wiley & Sons, 1972.

Lab view & Lab View for education 2010, by National Instruments, TX, 1986

Lewis, P. E. and J. P. Ward. *The Finite Element Method: Principles and Application.* Boston: Addison-Wesley, 1991.

LORD Sensing Systems. *LORD Sensing User Manual.* Vermont.

Martin, Harold and Graham F. Carey. *Introduction to Finite Element Analysis.* New York: McGraw-Hill, 1973.

McCormac, Jack C. *Structural Analysis,* Scranton, Pennsylvania, International Textbook Company, 1967

Meirovitch, Leonard. *Elements of Vibration Analysis.* New York: McGraw-Hill, 1975.

National Instruments Tutorial on Strain gauge , Application 078, 1998

Orfali, Robert and Dan Harkey. *Client/Server Programming with Java and CORBA.* New York: John Wiley & Sons, 1998.

Pohl, Ira. Object-Oriented Programming Using C++. London, England: Pearson, 1993.

Parallax devices from Parallax Development Corporation.

Reddy, J. N. *Introduction to the Finite Element Method.* New York: McGraw-Hill, 1984.

Rockey, K. C. et al. *The Finite Element Method: A Basic Introduction for Engineers.* New York: John Wiley & Sons, 1983.

Rogers, David E. and J. Allan Adams. *Mathematical Elements for Computer Graphics.* New York: McGraw-Hill, 1990.

Shigley Joseph E. and John Joseph Uicker Jr. *Theory of Machines and Mechanisms*. New York: McGraw-Hill, 1980.

Sedgewick, Robert. *Algorithms in C++ Data Structures with C & C++*. Boston: Addison-Wesley, 1992.

Strang, Gilbert. *Linear Algebra and Its Applications*. San Diago: Harcourt Brace Jovanovich, 1988.

"STEP" by Step Tools, Troy, NY.

Swaminathan, Prabhu. "Finite Element Analysis for Plane Stress and Strain and Plate Bending." Master's thesis, Northeastern University, 1986.

Swaminathan, Prabhu. "Mechanism of CAD Data Transfer between Various Organizations." *CAD Digest*. 2004.

Tong, Pin and John Rossettos. *Finite Element Method: Basic Technique and Implementation*. New York: Dover Publications, 1977.

Wang, Chi-Teh. *Applied Elasticity*. New York: McGraw-Hill, 1953.

Watt, Allan. 3D Graphics. New York: Addison-Wesley, 1993.

Weaver Jr., William and James M. Gere. *Matrix Analysis of Framed Structures*. New York: Van Nostrand Reinhold Co., 1965.

Weiskamp, Kieth, Loren Heiny, and Namir Clement Shammas. *Power Graphics Using Turbo C*. New York: John Wiley & Sons Inc., 1989.

Wilson, E. L. and Tetsuji Itoh. "An Eigen solution Strategy for Large Systems." *Computers and Structures* 16, no. 1–4 (1983): 259–265.

Zienkiewicz, O. C., R. L. Taylor, and J. Z. Zhu. *The Finite Element Method: Its Basis and Fundamentals*. New York: McGraw-Hill, 1967.

Other Sources of Reference

Fortran Compiler g77

XERCES DOM PARSER, open source

MySQL database

CATIA, a 3D tool from Dassault Systems

STAD III, and STRAP finite element software

Java programming language

Data structures in C, C++, and Java

Java and XML

UNIX operating systems, sockets, threads, client-server

Acknowledgements

Needless to say, I express my deep gratitude to the many contributors associated with this subject of CAD, CAE, and CAM.

I am also indebted to my many influential teachers and engineers who in many ways added to the skills that I possess to this day.

Index

www.ingramcontent.com/pod-product-compliance
Lightning Source LLC
Chambersburg PA
CBHW041634050326
40689CB00024B/4954